I0650346

John Malcolm Trout, Edward Trout

The Railways of Canada for 1870-71

John Malcolm Trout, Edward Trout

The Railways of Canada for 1870-71

ISBN/EAN: 9783744724203

Printed in Europe, USA, Canada, Australia, Japan

Cover: Foto ©berggeist007 / pixelio.de

More available books at **www.hansebooks.com**

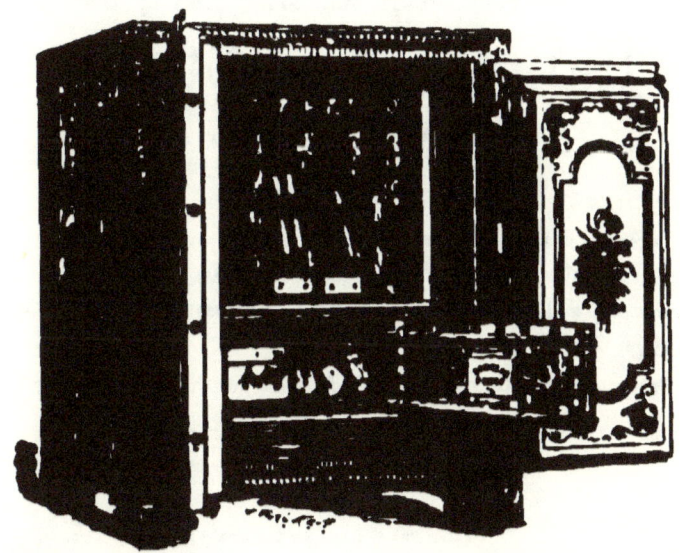

THE

RAILWAYS OF CANADA

FOR 1870-1,

SHEWING

THE PROGRESS, MILEAGE,

COST OF CONSTRUCTION,

THE STOCKS, BONDS, TRAFFIC,

EARNINGS, EXPENSES,

AND

Organization of the Railways of the Dominion.

ALSO,

A SKETCH OF THE DIFFICULTIES INCIDENT TO TRANSPORTATION IN CANADA IN THE PRE-RAILROAD DAYS.

BY

J. M. & EDWD. TROUT.

TORONTO:

PUBLISHED AT THE OFFICE OF THE MONETARY TIMES, NO. 60 CHURCH ST.

1871.

THE MONETARY

AND

Commercial Times

A WEEKLY NEWSPAPER,

DEVOTED TO

Finance, Commerce, Insurance, Railways, Mining, Investment

AND JOINT STOCK ENTERPRISE.

The only Journal in the Dominion which devotes

Special Attention to Railways.

Steps are being taken to render this feature more especially interesting and valuable, and for the better attainment of this object the co-operation of the Directors and Officers of the Railway Companies is cordially and earnestly invited.

News items relating to Railway Contracts, Official and other Changes and Improvements of every kind, are always gladly received and published. It is desired to make the railway department fully represent this great and growing interest.

Railway Managers Read it.

Railway Managers Subscribe for it.

Railway Managers Advertise in it.

Railway Managers Bind it for Reference.

It is invaluable to keep on fyle as a record of prices and of the commercial and general statistics of Canada.

SUBSCRIPTION PRICE.

Canadian subscribers	...	$2.00	per year.
British "	postpaid ...	10s.	"
American "	" .. gold	2.50	"

RATES OF ADVERTISING.

Casual Advertisements, each insertion. ..	10c.	per line.
Yearly Advertisements, per annum	$2.50	"

All letters should be addressed to

J. M. TROUT,

Business Manager.

Toronto, May 1st, 1871

INDEX TO RAILWAYS

INDEX TO ADVERTISEMENTS.

THE EARLY DAYS

OF

TRANSPORTATION IN CANADA.

The material progress of Canada has depended on nothing so much as the means of communication, the facilities for conveying men and goods.

On the discovery of Canada, when the whole country was covered with a primeval forest, the rivers and lakes formed the natural highways, the only means by which it was possible to travel. The birch bark canoe, which the Indians had from time immemorial used, had to be adopted by the first Europeans who made their way into the interior. When a fall or cataract was reached, the tiny vessel had to be hoisted on the shoulders of the travellers, and carried above or below the obstruction, together with whatever goods the party carried. Tents were generally out of the question; and the Jesuit missionaries frequently speak jocosely of having put up for the night at the sign of the moon; the stars their canopy, and chief or only covering. [Between Three Rivers and the country of the Hurons, on the east side of the Georgian Bay, which they named the Fresh Water Sea, and which the Indians called Attigouantan no less than forty *portages* had to be made]—that is, the canoe had to be taken out of the water and carried so many times—and the downward voyage, when sailing with the stream nearly all the way, consumed no less than thirty-five days, in which many perils to life and limb were encountered; a longer time than is now required to cross the continent five times from the Atlantic to the Pacific. The chief business of the country long centred in the fur trade, of which the beaver furnished the largest and most valuable supply. The boats used by the traders were necessarily limited in weight to what the voyageurs could carry on their shoulders over the portages. We are not going to waste time on a review of the fur trade or its progress, but it is worth while to note, as illustrating the inevitable slowness of the progress which it was possible to make in the absence of improved means of conveyance, that though Canada was discovered in 1514, the only means of getting into Lake Superior, possessed by the North-West Company, the most powerful organization that then existed in the country (the year 1800), was the bark canoe. It was large enough to carry eight or ten men, and a corresponding quantity of goods. It thus appears that for nearly three centuries the bark canoe, in one form or another, was the only reliance of Canadians, when extra long voyages had to be undertaken. On shorter voyages, other and superior craft were used.

At the close of the last century, it was the custom of Governor Simcoe to travel, from Kingston to Detroit, in a large bark canoe, rowed by twelve *chasseurs* of his own regiment; and followed by another boat, in which the tents and provisions were carried. The rule was to halt at noon for dinner, and in the evening to pitch the tents. When it was necessary to pass from one lake to the other—Ontario to Erie—by the portage at Queenston, this was then the only kind of vessel that could be used. On Lake Ontario he had the choice between the large bark canoe and a gun boat of eighty tons—that being the capacity of the Onondago—of which there were four. But only two of them, provided with sails and oars, were fit to carry either passengers or guns; and they were often pressed into the service of merchants, by whom either an equivalent in money was paid, or a return in like service in their vessels to the government was made.

(The cost of carriage, by every mode of conveyance then in use in the country was enormous. A bushel of Indian corn cost, by the time it reached Grand Portage, about thirty miles above Fort William, twenty shillings sterling; and Sir Alexander Mackenzie tells us it was the cheapest article of provisions the North-West Company could supply its men with, in the first year of this century. For the same sum ten bushels of corn can now be purchased in England, after having been carried a thousand miles in the interior of America and across the Atlantic.) But the North-West Company obtained the carriage of its stores very cheap, compared with what others paid. (The cost of carrying goods between Montreal and Kingston, before the Rideau or St. Lawrence canals were built, seems to this generation incredible) and is worthy of belief only, because it is stated on unimpeachable authority. Sir J. Murray stated, in the House of Commons, September 6, 1828, that, on a former occasion, the carriage of a twenty-four pound cannon cost between £150 and £200 sterling; that of a seventy-six cwt. anchor £676; and that when the Imperial Government sent out two vessels in frames, one of them, a brig, cost the country in carriage, the short distance between these two cities, the enormous sum of thirty thousand pounds sterling; nearly one hundred and fifty thousand dollars. The same service could now be performed for a mere trifle. In the early days of the Talbot settlement—about 1817—so called from a large district of country in Western Canada having been granted to Col. Talbot to place settlers upon, we have the authority of Mr. Edward Ermatinger, the biographer of that eccentric pioneer, for the statement that eighteen bushels of wheat were required to pay for a barrel of salt, and that one bushel of wheat would no more than buy a yard of cotton. From the difficulty of getting seed grain over the wretched roads of this new country, the struggling pioneer sometimes had to pay as high as two dollars a bushel for wheat, which sold in other parts of the province, where communications were better, for about three shillings and three pence a bushel, and other things necessary to his comfort and subsistence were proportionately dear.

The enormous rates of Atlantic freights, in those early days, show the immense improvements that have since taken place in ocean navigation. Mr. David Anderson, who, in 1814, published a book to prove the importance of the British American Colonies to England, estimated the freight of a quantity of wheat sufficient to make a barrel of flour, from Canada to England, at a pound sterling, nearly five dollars. He was obliged to make an estimate, when dealing with a barrel of flour, because "breadstuffs" were then shipped to England only in their unground state; and if his figures be reliable, Atlantic freights on this form of "the staff of life," were seven times as high as at present. We suspect, however, that his estimate was too high.

The average cost of freight on all the grain taken to England is added to the price of the grain; and if it costs five or six times as much to take grain to that market from one country as it can be taken for from another, the producer in the former country is at a great disadvantage in the competition he is obliged to meet. Discriminating duties could not be expected to make up the difference. Lying under these enormous disabilities, in respect to the transmission of produce from the place of production to the ultimate market, it was inevitable that the exports of Canada in grain should be low. In the quarter of a century ending with 1824, when the practice of grinding wheat for exportation had begun, Canada had exported only 563,212 bbls. of flour, and 4,833,190 bushels of wheat. Her population was small; but the growth of population under this condition of things must necessarily be the reverse of rapid.

Between Quebec and Montreal, and on Lake Ontario an improved kind of craft was used long before the same thing was possible between Montreal and Kingston. In 1795, three small merchant vessels, owned at Kingston, used to make eleven voyages a year to the portage at Queenston; they formed the bridge between Kingston and Queenston; and long after, so little was foreseen of the future tracks of commerce, it was thought that the latter place would always continue to play an important part in the trade of the country. These vessels were, probably, from fifty to two hundred tons burthen, as Weld tells us, there were merchant vessels of that class on the lake at that date. Canoes and batteaux were also much used; all the coasters on the American side being of the latter class. Nearly all the British commerce of the lake was between Kingston and Queenston. The vessels seldom called at any other point. The number of vessels must have been small; for, if we may trust a statement published in the newspapers of the time, there were, in 1812, seventeen years after, on the Canadian side of Lake Ontario, only three vessels of over forty tons each. In 1826, in spite of the war that had intervened; the number of vessels of that size had increased to between thirty and forty, and some reached nearly, or quite, one hundred tons. At the former date, 1795, the fare between Kingston and Niagara was ten dollars, first class, and half that sum second class. The freight on goods between Kingston and Queenston was about nine dollars a ton (thirty six shillings sterling) nearly as much as would have been paid for carrying them across the Atlantic, before the war then raging in Europe broke out. But ships were costly to construct, and wore out rapidly; sailors had to be brought up from the ocean, and retained on pay during the five or six winter months when the harbors were frozen up. Ship carpenters, brought from the States, worked in summer and returned home in winter. Added to this rate of freight was the previous carriage, sometimes of over two thousand miles, inland, before they were put on board at Queenston portage. Over this portage, sixty wagons would sometimes pass in a day. The upper landing place was on Chippawa Creek. Merchandize took this route westward by

Detroit to Michilmackinac, and beyond. This portage trade gave the same importance to Queenston that Lachine received from a similar kind of traffic.

The first steamboat that ran between Quebec and Montreal appears to have been built in 1811, by Mr. John Molson, well known as the father of steamboat enterprise on the St. Lawrence. We find by the journals of Lower Canada that a bill was brought in, in that year, to grant him the exclusive right of navigating with one or more steamboats that part of the river; but though it passed through committee, it did not become law. Next year it was again introduced on petition. The petition sets forth that Mr. Molson had already built a steamboat, at great expense, which would afford the means, at a small cost to the public, of a speedy and convenient passage between the two cities; the only means of making it then in use being "fatiguing from the nature of the vehicle, and inconvenient both for lodging and nourishment." The petition did not mention the number of years during which this exclusive privilege was desired. The Legislative Council passed the bill, and inserted the term of fourteen years; but when it came before the Assembly, in Committee, the House was counted out for want of a quorum, only thirteen members being present, among them L. J. Papineau, who was favourable to the measure. Nevertheless, steamboat communication was established on that part of the St. Lawrence, through the enterprise of Mr. Molson. It lessened the cost, shortened the time, and banished many of the discomforts of travelling between the two chief cities of Lower Canada.

Twelve years later, there were no less than seven steamboats plying between Quebec and Montreal. Five of them appeared in Edward Allen Talbot's eyes nearly as long each as a forty gun frigate. The double row of sleeping berths, on each side of the cabin, were thought to be surpassing luxuries, where state-rooms were unknown; though they would now fail to command any but second class passengers. And the charge, £3 sterling, over fourteen dollars and a half from Quebec to Montreal, and ten shillings less the other way, would now take a passenger all the way from Hamilton to the Saguenay by steamboat, and from Sarnia to Portland by rail. But the rates of passage were soon reduced, by the natural operation of competition, to a moderate figure. By the year 1829, deck passage on these steamers could be had for a dollar and a half; and a passage could be had on such conveyance as then existed, from Montreal to Kingston, for five dollars more.

Upper Canada was only a little later in availing itself of the facilities of steamboat navigation. The Frontenac, the first Lake Ontario steamer, was not built till 1816. She cost £15,000, which is nearly three times as much as any other boat on that lake cost for the next decade, as the following figures, which represent the commercial steam marine of Lake Ontario in 1826, show:

NAMES OF STEAMERS.	COST.
Frontenac	£15,000
Queenston (estimated)	5,000
Niagara	6,000
Charlotte	3,500
Toronto	2,500
Canada	5,000
Dalhousie	2,500
Total	£39,500

The Frontenac, Howison tells us, was the largest steamboat in Canada; her deck being seventy-two feet long and thirty-two feet wide; seven hundred and forty tons burthen, and drawing eight feet of water. The time has long since passed when any one would think of using, on these waters, so small a steamer for passenger traffic. But the size of Canadian steamers soon underwent an increase. In 1829, the Lady Sherlock, which ran between Quebec and Montreal, was one hundred and forty-five feet long, and the Chambly was only three feet shorter. Before the Lachine Canal was built small steamers managed to stem the Lachine rapid, which they overcame by going obliquely against the current and taking advantage of the side eddies.

It is curious to note that, at a distance of about five years, Upper Canada followed Lower in the inauguration of steamboat enterprize; and that she counted seven steamboats on Lake Ontario two years after Lower Canada hed placed that number between Quebec and Montreal. The fare charged by the first Upper Canada steamboat was twelve dollars from Prescott to Toronto, and half as much again to Hamilton.

But while these two sections were provided with steamboat accommodation, the intermediate distance between Kingston and Montreal was still, on account of the interruptions occasioned by the rapids, obliged to content itself with more primitive modes of communication.

The flat bottomed *Batteaux*, made of pine boards, and narrowed at bow and stern, forty feet by six, with a crew of four men and a pilot, provided with, oars, sails and iron shod poles for pushing, continued to carry, in cargoes of five tons, all the merchandise that passed to Upper Canada. Sometimes these boats were provided with a makeshift upper cabin, which consisted of an awning of oilcloth supported on hoops like the roof of an American, Quaker or Gipsey wagon: provided with half a dozen chairs and a table, this cabin was deemed the height of primitive luxury. The Batteaux went in brigades, which generally consisted of five boats. Against the swiftest currents and rapids, the men poled their way up; and when the resisting element was too much for their strength, they fastened a rope to the bow, and plunging into the water, dragged her by main strength up the boiling cataract. From Lachine to Kingston, the average voyage was ten or twelve days; though it was occasionally made in seven; an average as long as a voyage across the Atlantic now. The nature of the route over which they travelled had dictated the construction of these boats; the main object being that they should draw as little water as possible. A Batteaux of two tons, if heavily laden, had to be lightened to pass over the Long Sault, when the water was low.

The Durham boat, also then doing duty on this route, was a flat bottomed barge; but it differed from the batteaux in having a slip keel and nearly twice its capacity.

This primitive mode of travelling had its poetic side. Amid all the hardships of their vocation, the French Canadian boatmen were ever light of spirit, and they enlivened the passage by carrolling their boat songs; one of which inspired Moore to write his immortal ballad, better known among the generality of English readers than those of the French that preceded it.

The loss of time, from the slowness of the old modes of travel, was a very serious matter. Edward Allen Talbot, who published a book on Canada, in 1824, has some facts cited from his own experience on this point. We should be sorry to guarantee the general accuracy of this prejudiced and splenetic work; but the author may be trusted when he tells us that himself, his father, and the rest of the family were thirteen days in a Durham boat, between Lachine and Prescott. To the loss of time by this mode of travelling was added the discomfort arising from a part of the passengers having to sleep at night, when the boat came to a stand, in the open air, on shore; the wretched little cabin—not of the awning kind, it is presumed—not being sufficient to accommodate a single family.

The dangers of this mode of travelling, like that by canoe which it had superseded, were very great; those of the Long Sault being especially dreaded. Mr. Boulton, in his topographical description of Upper Canada, published in London, in 1824, says: "Boats may pass near shore, but where misfortune has driven either a boat or a raft into the strong part of the current, it hath seldom happened that a life has been saved. A melancholy instance of the danger of this just occurred in the late French war, when several boats and their crews were entirely lost." But familiarity with the currents had reduced the danger to a minimum; and the surplus grain of Upper Canada was now taken down on rafts or in boats, with a great degree of safety. Attempts had been made to take lumber down from the most distant points on Lake Ontario; but Mr. Boulton conceived "the risk to be far above the probable advantage;" a risk which, in these later days, we have learnt to count very little.

As between the Batteaux and the Durham boat, the balance of safety lay on the side of the former. An example from the experience of Isaac Weld, the traveller, when passing from Montreal to Quebec, in the summer of 1795, will show this in a striking manner. After leaving Montreal, "we had," he says, "reached a wide part of the river, and were sailing under a favourable wind, when suddenly the horizon grew very dark, and a dreadful storm arose, accompanied by loud peals of thunder and a torrent of rain. Before the sail could be taken in, the ropes which held it were snapped in pieces, and the waves began to dash over the sides of the batteaux, though the water had been quite smooth five minutes before. It was impossible now to counteract the force of the wind with oars, and the batteaux was consequently driven on shore, and the bottom of it being quite flat, it was carried smoothly upon the beach without sustaining any injury, and the men leaping out of it drew it on dry land, where we remained out of all danger till the storm was over. A keel boat, however, of the same size, could not have approached nearer to the shore than thirty feet, and there it would have stuck fast in the sand, and probably have been filled with water."

The great leading roads of the Province had received little improvement beyond being graded, and the swamps made passable by laying the round trunks of trees, side by side across the roadway. Their supposed resemblance to the King's corduroy cloth, gained for these crossways the name of corduroy roads. The earth roads were passably good only when covered with the snows of winter, or dried up with the summer sun; and even then a thaw or a rain made them all but impassable. The rains of autumn, and the thaws of spring, converted them into a mass of liquid mud, such as amphibious animals might delight to revel in. Except an occasional legislative grant of a few thousand pounds for the whole Province, which was ill expended, and often not accounted for at all, the great leading roads, as well as all other roads, depended, in Upper Canada, for their improvement on statute labour. In 1831, every male inhabitant not rated on the assessment roll, was liable to two days labour on the roads; a person rated at not more than twenty-five pounds, to three days labour; if over fifty, and less than seventy-five, four days; at one hundred pounds five days; at two hundred pounds, seven days; at three hundred, nine days; at four hundred, eleven days; at five hundred twelve days. This labor was languidly performed, or, when possible, evaded altogether; substitutes were difficult to get, and money to pay them with equally so. In that year, £20,000 was granted by the Legislature for the improvement of roads; and Mr. Ruttan, in a pamphlet published the next year, stated that £9,000 of it remained unaccounted for. In 1835, no less a sum than £50,000 was granted for the improvement of roads; but this sum, even if economically expended, would go a very little way in forming good roads, over distances that embraced many hundreds of miles. In 1836-7, a Session of recklessly improvident grants of all kinds, £500,000 was authorised to be raised for roads; but it was of no more value than the several other similar authorisations, amounting in the aggregate to several millions of dollars, when the credit of the Province was at zero, and its whole revenue was not one-third as much as that of one of our richest municipalities to-day.

At the time of the union, in 1841, the whole revenue of the Province was only £78,000; that of Toronto was, in 1870, $1,362,169 25. Formerly the small grants for this purpose were jobbed and squandered by members of the Legislature, under a system in which no one was responsible, and every member could propose a money grant without the previous authority of the Crown. In 1840, Chief Justice Robinson estimated the whole amount that had been expended on Macadamized roads, in Upper Canada, at £200,000—$800,000. After the union, a large portion of the Imperial guaranteed loan of £1,500,000, was expended on this kind of roads; but the money was so distributed that the great leading routes were seldom more than partially improved.]

The only road on which it was possible, in 1837, to take a drive, near Toronto, was Yonge Street, which was Macadamized a distance of twelve miles. Mrs. Jamieson describes the Canadian stage coach as being, at that time, like the American, a "heavy lumbering vehicle, well calculated to live in roads where any decent carriage must needs founder." These were the better sort, on the great roads. Another kind were "large oblong wooden boxes, formed of a few planks nailed together, and placed on wheels, in which you enter by the window, there being no door to open or shut, and no springs." On two or three wooden seats, suspended on leather straps, the passengers were perched. The behaviour of the better sort, in a journey from Niagara to Hamilton, is described by this writer as consisting of a "reeling and tumbling along the detestable road, pitching like a scow among the breakers of a lake storm." The road was knee-deep in mud, "the forest on either side dark grim and impenetrable."

Bad as this was, there were men scarce past the prime of life, who, contrasting it with their recollections and experience, might be excused for thinking it a very acceptable mode of travelling. They could remember the time when it was impossible to thread their way among the stumps of trees and fallen timber that encumbered the road, with a rude cart and a yoke of oxen; when the Duke de la Rochfoucalt Liancourt, in 1795, described this very road as one of the worst he had seen in America; when it was passable only on horseback, and then, he tells us, "but for our finding now and then some trunks of trees in the swampy places, we should not have been able to disengage ourselves from the morass." Thirty years latter, Mr. Wm. L. Mackenzie described the road between Toronto and Kingston, as among the worst that human foot ever trod. And down to the latest day before the railroad era, the travellers in the Canadian stage coach were lucky if, when a hill had to be ascended or a bad spot passed, they had not to alight and trudge ancle deep through the mud.

In Lower Canada the *Maîtres* and *Aides de Poste* formerly kept conveyances for the carriage of passengers at stated post houses; and the rates of charge were fixed by law. They received ten-pence a league for a horse and cart or sleigh, or for a horse and harness without either, for conveying a weight of six hundred pounds, and four-pence for every additional horse, conveying a weight of one thousand pounds; and seven-pence half-penny a league for a saddle-horse. The Act establishing these post houses having expired, the *ci-devant Maîtres* and *Aides de Poste*, petitioned for their re-establishment, with a legalized tariff, in 1812. But a committee to whom the petition was referred, reported adversely; and thenceforth the carrying of passengers on land seems to have been left to the natural law of competition.

The rate which it was possible to travel in stage coaches, depended on the elements. In spring, when the roads were water-choked, and rut-galled, the rate might be reduced to two miles an hour, for several miles on the worst sections. The coaches were liable to become embedded in the mud, and the passengers had to dismount and assist in prying them out by means of rails obtained from the fences. Various forms of accidents occurred, and the total percentage was probably not less than fifty per cent. more than on railways at present. The cost of travelling, in fares, to say nothing of time and expenses on the way, where the driver was generally in league with the tavern-keepers, by whom he was used as a decoy, was nearly three times what it is on railways. In the dry weather of summer, and the snows of winter, the worst roads became tolerably good; and stories of incredible speed being made, in sleighing, are still told. It is alleged that Mr. Weller—the immortal stage-coach owner—once drove Lord Sydenham from Toronto to Montreal, by means of successive relays of horses, in twenty-six hours; and a story is told of a still more surprising feat being performed, in the same way, between Portland and Montreal. It was a race between Boston and Portland, which could carry the English mail most rapidly to Montreal. The Portland party made the distance, which is nearly three hundred miles, in twenty hours. The result of this contest is said to have been one of the causes that led to the adoption of Portland as the terminus of the railway from Montreal, instead of Boston. But these exceptional cases prove nothing in favour of a mode of travelling, which, taken altogether, in the varying seasons, was tedious and uncomfortable, and involved an outlay of time and money that would now be thought unendurable.

We have said enough to prove the proposition with which we set out, that the material progress of Canada depends, and has always depended, more upon the facilities for communication than anything else. This brief retrospect will give the present generation some adequate idea of the advantages it possesses over those that went before; a kind of knowledge which may, if rightly used, be turned to practical account.

PROGRESS OF RAILWAY CONSTRUCTION.

A good idea may be formed of the great relative importance of our railways, in their bearing upon the financial and industrial interests of the Dominion, from the fact that their annual receipts are nearly equal to the entire public revenue, or about FOURTEEN MILLIONS OF DOLLARS.

To a country with the physical configuration of the Dominion—stretching from the Atlantic to the Pacific, and settled only on a relatively narrow frontier strip—cheap and rapid communication is one of the first requisites. The diversified products of the eastern and western sections require to be constantly interchanged in order to meet the wants of both. And nothing will so powerfully tend to consummate the great object aimed at in forming our Confederate Constitution—the real and lasting union of the people of all these provinces—as supplying the best possible facilities for the interchange, not merely of commodities, but of thought, by the means of correspondence and personal intercourse. The Intercolonial was no doubt projected, more as a political than as a commercial undertaking, and very great advantages may be expected from it in the way of bringing about acquaintanceship, creating and riveting social ties and commercial relations, breaking down antipathies and creating the sense of a common interest. Let us hope that as a military convenience it will never be called into requisition. The same necessity that forced the construction of the Intercolonial operates to urge the building of a CANADIAN PACIFIC LINE, which, great as the undertaking is, will undoubtedly be proceeded with without any unnecessary delay. These two lines, when completed, will, with our other great public work, the Grand Trunk Railway, extend as a vast iron girth across the Continent, forming a grand National Highway of three thousand miles in length, or in all, six thousand continuous miles of railway track.

The brilliant success of Mr. George Stephenson's engine "Rocket," on the Liverpool and Manchester Railway, drew the attention of the world to this new and marvellous triumph of genius. The £500 prize offered by that Company was won by the engine named—the trial taking place on the 6th October, 1829. This engine, which weighed four tons, made on the level, with 12¾ tons attached, 29½ miles per hour. A result so astounding to the ideas of our ancestors, who regarded any means of travel faster than a stage coach at ten miles an hour as tempting Providence, was soon published far and near. In spite of the most unscrupulous and persistent opposition, this innovation forced its way into public notice. Railways soon became what they now are, one of the most marked characteristics of our modern civilization.

As a means of opening up a new country for settlement, railways are incomparably the best and most effective, viewed in the light of results, that human skill has yet devised. Like the arteries and veins in the human body, they are the channels which vitalize the extremities of a country, and bring them into direct and immediate connection with the centres of commerce. They give value to natural products before valueless, because out of the reach of consumers; change sterility into productiveness; convert the wilderness into cultivated farms, as if by magic, and substitute for the profitless hunting of the wild man of the forest, the peaceful and remunerative operations of modern husbandry. Railways have accomplished all this in Canada, but the work has only fairly begun.

Very soon after the first railways were commenced in Great Britain and in the United States, several projects were formed and discussed for the construction of lines in Canada. From 1832 to 1840 a large number of charters were obtained in all the Provinces, but the great majority of the schemes so authorized proved abortive, and the Acts suffered to remain on the statute book as a dead letter. A full list of these charters with a brief sketch of their principal features is given in another place.

In 1836 the first attempt at working a railway in Canada was made. The St. Lawrence and Champlain, (now the Montreal and Champlain,) was opened in that year; the rails were of wood with flat bars of iron spiked on them, and from the tendency of this class of rail to curl or bend upward as the wheels passed over it, it became known as the "snake rail." From this awkward peculiarity it often happened that the rails came into contact with the body of the cars or other rolling stock, in which case both fared badly. The first locomotive used on the Line was sent from Europe, accompanied by an engineer, who for some unexplained reason had it caged up and secreted from public view. The trial trip was made by moonlight in the presence of a few interested parties, and it is not described as a success. Several attempts were made to get the "Kitten"—for such was the nick-name applied to this pioneer locomotive—to run to St. Johns, but in vain; the engine proved refractory and horses were substituted for it. It is related, however, that a practical engineer being called in from the United States, the engine which was thought to be hopelessly unmanageable, was pronounced in good order requiring only "plenty of wood and water." This opinion proved correct, for after a little practice the "extraordinary" rate of speed of twenty miles per hour was attained. Other difficulties were soon overcome and the first Canadian railway became an accomplished fact.

The first locomotives used in Canada and the first sent across the Atlantic to British North America were the "James Ferrier," "the Montreal" and the "John Molson." They were built by Messrs. Kinmond & Co., of Dundee, Scotland, in 1847, and shipped in the spring of 1848. The first two were used on the Montreal and Lachine railway, and the third ran from St. Lambert to St. Johns on the Montreal and Champlain railway. Some of them are still running.

It was fully a decade subsequent to the date of the opening of the St. Lawrence and Champlain Railways that the Huron and Ontario and Great Western projects took practical shape in Upper Canada, although charter powers were conferred for the construction of the former line as early as 1833 and for the latter in 1834. So little was the progress made that in 1850 there were but fifty-five miles of railway in all the Provinces.

In 1849 a general Act was passed known as the "Guarantee Act" which empowered the Government to aid any railway not less than seventy miles in length by guaranteeing the payment of six per cent interest on a sum not to exceed one half the total cost of the road. In 1858 the Government guarantee was extended to the principal, the Government taking a first lien on the railways so aided. Though this policy never realized the anticipations formed of it, yet it had the effect of giving a powerful stimulus to railway enterprise. Then commenced the first railway era in which all our present lines were constructed. The Grand Trunk Railway Company was incorporated in 1852 and the work pushed to its final completion in 1857.

We shall not enter upon the details relating to the different lines here, but merely present the subjoined table which is compiled from official sources and shews the dates at which the different sections of the various lines were completed, being an epitome of the progress made in railway construction to the present time:—

RAILWAYS OF CANADA.

DATE OF OPENING EACH SECTION.

NAMES.	NAME OF SECTION.	Date of Opening.	Length of Sections.	Total Length.
Great Western	Susp. Bridge to Hamilton	Nov. 10, [illegible]	43	
"	Hamilton to London......	Dec. 21, [illegible]	76	
"	London to Windsor ...	Jan. 27, 1854	110	
"	Harrisburg to Galt branch	Aug 21, 1854	12	
"	Galt to Guelph branch...	Sep. [illegible], 1857	15	
"	Hamilton to Toronto b'h	Dec. 3, [illegible]	[illegible]	
"	Komoka to Sarnia........	Dec. 27, [illegible]	51	
"	Petrolia & Kerth Branches		15	[illegible]
Grand Trunk	Toronto to Guelph	July, [illegible]	50	
"	Guelph to Stratford	Nov. 17, [illegible]	[illegible]	
"	Stratford to London	Sep. 27, [illegible]	31	
"	St. Mary's to Sarnia	Nov. 21, [illegible]	70	
"	Toronto to Oshawa	August, [illegible]	[illegible]	
"	Oshawa to Brockville ...	Oct. 27, [illegible]	175	
"	Brockville to Montreal ..	Nov. [illegible], [illegible]	125	
"	Victoria B. & approaches	Dec. 16, [illegible]	6	
"	Montreal to St. Hyacinthe	[illegible] 1847	[illegible]	
"	St. Hyacinthe to Sherb'ke	August, [illegible]	[illegible]	
"	Sherb'ke to Province Line	July, [illegible]	[illegible]	
"	Richmond to Quebec ..	Nov. 27, [illegible]	[illegible]	
"	Chaudiere Junction to St. Thomas	Dec. [illegible], [illegible]	41	
"	St. Thomas to St. Paschal	Dec. 31, [illegible]	[illegible]	
"	St. Paschal to Rivière du Loup	July 2, [illegible]	[illegible]	
"	Kingston branch	Nov. 10, [illegible]	2	[illegible]
Northern................	Toronto to Bradford	June 13, [illegible]	42	
"	Bradford to Barrie...... .	Oct. 11, [illegible]	[illegible]	
"	Barrie to Collingwood ...	Jan. 1, [illegible]	[illegible]	
"	Bell Ewart branch. .		1 24	
"	Barrie branch	[illegible]	1.[illegible]	[illegible]
Buffalo & Lake Huron,	Fort Erie to Paris	Nov. 1, [illegible]	[illegible]	
" "	Paris to Stratford	Dec. 22, [illegible]	[illegible]	
" "	Stratford to Goderich ...	June 28, [illegible]	45	
" "	From temporary terminus to Station East St.	May 16, [illegible]	1.27	[illegible]
London & Pt. Stanley.	Lake Erie to London ..	Oct. 1, [illegible]	[illegible]	
Erie & Ontario	Lake Ontario to Chippawa	July 3, 1854	17	
Ottawa & Prescott......	From the St. Lawrence to Ottawa City......... .	Dec. 1854	54	
Montreal & Champlain	Montreal to Lachine......	Nov. 1847	8	
" "	Caughnawaga to Mooers' Junction...	Aug. [illegible]	32	
" "	St. Lambert to St. Johns (old portion July 1836)	Jan. [illegible]	20	

RAILWAYS OF CANADA—*Continued.*

NAMES.	NAME OF SECTION.	Date of Opening.	Length of Section.	Total Length.
Montreal & Champlain	St. Johns to Rouse's Pt.	Aug. 1851	21.76	
				177.76
Carillon & Grenville ...	....	Oct. 1854		12.75
St. Lawrence & Industry	Lanoraie to St. Industrie	May 1850		12
Port Hope, Lindsay & Beaverton	Port Hope to Lindsay ..	Dec. 30, 1857	43	
Do., do.	Millbrook to Peterboro B	Aug. 18, 1858	23	
Do., do.	Lindsay to Beaverton ...	Jan. 1870	23	
				89
Welland	Port Dalhousie to Port Colborne	June 27, 1859		25
Brockville & Ottawa...	Brockville to Almonte ...	Feb17, Aug 1859	51.25	
" " ...	Smith's Falls to Perth b'h	Feb. 17, 1859	11.54	
" " ..	Tunnel from temporary Station to Harbor	Dec. 31, 1860	.75	
				63.54
Stanstead, Shefford & Chambly........	St. Johns to W. Farnham	Jan. 1, 1859	13	
Do., do.	W. Farnham to Granby.,	Dec. 31, 1859	15	
				28
Cobourg & Peterboro..	Cobourg to Harwood.. ...	May 1854	14	
" "	Junction to Ore Bridge...		9	
				23
Nova Scotia........... .	To Mile House	Feb. 1855	4	
"	Mile House to Bedford...	July 1855	8	
"	Bedford to Grand Lake...	Jan. 1857	23	
"	Grand Lake to Elmsdale.	Jan. 1858	30	
"	Elmsdale to Shubenacadie	March 1858	39	
"	Shubenacadie to Truro...	Dec. 15, 1858	61	
"	Truro to Pictou	May 31, 1867	52	217
"				
Windsor Branch ...	Junction to Windsor......	June 3, 1858	32	32
Windsor & Annapolis..	Wi'dsor to Annapolis N.S		85	85
Wellington, Grey & Bruce	To Elora	Sep. 15, 1870	16	
Do., do.	To Alma	Dec. 1870	5	
				21
N. Brunswick & Canada	St. Andrews to Barber Dam	Oct. 1, 1857	34	
" "	Barber Dam to Canterb'y	Dec. 1858	31	
" "	St. Stephens Branch......		19	
" "	Woodstock "		11	
" "	To Richmond..............	July 1862	23	
" "	Houlton Branch...........		8	
				126
European & N American	Western Extension—Fairville to St. Croix ..	Dec. 1869	88	
" "	Fredrickton Road.........		22¾	
" "	Eastern Extension.........		36½	
				147.25
Quebec & Gosford	Quebec to Gosford.........	Dec. 1870	26	26
Add for lines under	construction...............			100
	Total.........	....	... ...	2,779.10

For two years succeeding the completion of the Grand Trunk, which also marked the date of the most terrible commercial crisis through which this country ever passed, nothing was done; railway enterprise was paralyzed; the roads being operated did not prove remunerative, dispelling in the rudest fashion the fallacious hopes raised by their too sanguine promoters; the lines fell into serious pecuniary difficulties. The Northern got into so low water as to be seized by the Government for delinquencies with respect to the public lien. It became apparent that the figures paid for construction were extravagant; that the money which should have served for an ample equipment was lavishly disbursed on the permanent way, leaving the leading lines in anything else than a prosperous condition. These circumstances, together with the complete prostration which overtook every industry and every interest in the country, directed a strong public prejudice against railways, and effectually stamped out for the time all railway progress.

The ten years from 1860 to 1870 furnished ample time for reflection on the errors of the past, and recuperation from the disastrous collapse of the speculative period named; and 1870 witnessed a complete revival of railway enterprise, modified and restrained by the lessons of past experience.

It rests with us now to see that none but legitimate projects are encouraged, if the efforts being made are to result to the profit of the country and to the credit of this most useful class of our public undertakings.

RAILWAY FINANCE.

When the first railway charters were granted in Canada, there seems to have been a notion that the companies would be likely to make too much profit, and that their earnings over and above a fixed dividend formed a fair subject of taxation. More than one charter provided that a moiety of the net earnings over a dividend of ten per cent., which should have been paid ever since the work of construction commenced, should go into the public exchequer. In other cases the tariff was to be regulated by the amount of yearly dividend. It is not necessary to say that these clauses never became operative. The following embrace the various phases of railway finance which have been resorted to in Canada :

1. Authority given to Government to issue debentures by way of loan to railway companies. This authority was given long before any company was in a position to avail itself of the offer.

2. Authority to grant a like loan with a provision that if the company did not pay the interest on the Government debentures, the property of the cities and townships benefited should be assessed for the same.

3. Government guarantee of the interest on railway companies' bonds.

4. Government guarantee of railway companies' bonds, as well as the interest thereon.

5. Direct issue of Government bonds to railway companies, with a first mortgage on the property of the companies as security.

6. Government guarantee of share capital (asked but not granted.)

7. Municipal loans to railway companies.

8. Municipal subscription to railway stock.

9. Municipal bonuses to railway companies.

10. Government bonuses to railway companies.

11. Raising capital by lottery; authorized but not carried out.

12. Imperial Government guarantee of capital with which to construct the Intercolonial Railway.

13. Share capital, locally contributed, and issue of bonds.

14. Share capital, chiefly English, combined with Government aid, in some of the forms above mentioned, and issue of various degrees of bonds, under different names.

15. Aid in the shape of lands through which the road would run.

16. Practical release of Government guarantee, by placing it hopelessly behind other claims, after railway companies became embarrassed.

17. Composition of Government claim accepted, when railway companies became embarrassed.

18. Assumption of liabilities incurred by municipalities in aid of railroads by the Government, the Government becoming the creditor of the municipalities.

19. Direct construction of railways by Government.

A marked peculiarity in the mode of financing adopted by recent projects consists in seeking municipal and government aid in the shape of bonuses and not as subscribed stock. This is the most honest and proper method, in that it raises no false hopes and prevents the disappointment and embarrassment that would certainly otherwise ensue. Many of the municipalities have contributed most liberally in this form; the Government of Ontario, with commendable wisdom, have set aside $1,500,000 to be granted by way of bonus, in sums ranging from $2,000 to $4,000 per mile, according to circumstances. In Quebec, aid has been granted to the Northern Colonization Railway, to run from Montreal to Ottawa City, in the shape of lands to the extent of 1,200,000 acres, for 120 miles of road; they have granted 800,000 acres to the proposed St. John (N. B.) and River Du Loup Railway, and 2,000,000 to the North Shore Railway, to run from Quebec to Montreal, on the North Shore of the St. Lawrence. The Dominion Government in addition to building the Intercolonial of 500 miles in length at an estimated cost of $20,000,000, has also agreed to commence within two years the Canadian Pacific Railway, three thousand miles in length, to cost one hundred millions of dollars.

AMERICAN RAILWAYS.

Immediately after the results of the trial of Mr. Geo. Stephenson's Engine on the Liverpool and Manchester Railway, a most important agitation sprang up in the United

States. A section of 14 miles of the Baltimore and Ohio Railway was completed in 1830, and opened for traffic. It was worked by horse-power. In the next year a locomotive engine, the first of American manufacture, was placed on this line. In the same year an English engine, weighing six tons was obtained for the Mohawk and Hudson, but this proving destructive to the permanent way, an engine of American make, weighing only three tons was substituted in its place. In 1832, the South Carolina Railway was opened, also the New York and Harlem, and the Camden and Amboy, in New Jersey. The Boston and Lowell, in the State of Massachusetts, was commenced in 1831, and the Boston, and Providence and Boston, and Worcester, in the following year these three roads were completed in 1835. The Newcastle and Frenchtown, extending from Chesapeake to Delaware Bay was commenced in 1831 and finished in 1832. All these schemes were crude and ill-judged. As in Canada, the estimates always fell far short of the actual cost. This, with the defective character of the works rendering constant repairs necessary, sadly embarrassed nearly every enterprise undertaken. The railways did not prove remunerative and became a serious burden on the capital and industry of the country; a state of affairs which brought about those wide-spread failures, and sweeping financial disasters, known in the aggregate as the crisis of 1837. This collapse gave the quietus to railway enterprise for a period of at least ten years. Many projects on which a good deal of money had been spent were wholly abandoned; others were gone on with. But the total miles constructed in the ten years following would scarcely equal the number completed in a single year since. From the small beginnings of forty years ago, the railway interest in the United States has grown enormously; the total mileage is now 50,000 in round number, and these are being added to at the rate of 3,000 to 4,000 miles of new lines annually.

The liberal public policy of the United States Govern

ment with reference to this class of public works has had much to do with their almost marvellous expansion, and with the equally marvellous results that have followed in the developement and progress of the country. It is estimated that the total amount invested in American railways approximates very closely to *two thousand millions* of dollars! The roads did not cost even three-quarters of this sum (which represents their capital accounts), the difference of over one quarter being made up by the process known as "watering." By this means the capital of eighteen railways was increased in five years by the sum of $322,091,858, eight of which more than doubled their entire capital in that time. This practice, which originated first in a shrewd business move to benefit the stockholders without giving the appearance of distributing extravagant profits, has developed into a huge iniquity, which ought, in the interests of the public, to be restrained, if possible, by legislative enactment. Anyone who has watched the unprincipled operations of the Erie Railway managers, which have become a world-wide scandal, will see the necessity of effective checks being imposed upon corporations wielding so much power for good or evil. The extent of this power can be easily appreciated when it is remembered that the annual earnings of all the United States roads now exceed four hundred millions of dollars, or nearly twelve dollars per head for the entire population.

In addition to the grant of thirty-five millions of acres of public lands to the Pacific Railway, already constructed, the United States Government issued $63,616,000 in 6 per cent currency bonds in aid of that undertaking. The whole line is 3,300 miles in length, from the Atlantic to the Pacific Ocean. The public aid was, however, only extended to 2,500 miles of the railway. The bonds were issued upon 300 miles at the rate of $48,000 per mile, upon 976 miles at the rate of $32,000 per mile, and upon 1244 miles at the rate of $16,000 per mile. A second mortgage was accepted by the

Government as security for the loan, and the companies were authorized to issue their own bonds to an amount equal to the Government subsidy, the same being made a first mortgage over the whole of the companies' effects. The annual interest on the subsidy is $3,934,560.

Subjoined is a statement of the amount of lands granted by Congress to the States named for the construction of railways up to the 1st July, 1869.

STATES.	ACRES GRANTED.
Illinois	2,595,053
Mississippi	2,062,240
Alabama	3,729,120
Florida	2,360,114
Louisiana	1,578,720
Arkansas	4,744,272
Missouri	3,745,160
Iowa	7,331,208
Michigan	5,327,931
Wisconsin	5,378,360
Minnesota	7,783,403
Kansas	7,753,000
California	2,060,000
Oregon	1,660,000
Total	58,108,581

	ACRES.	
Grant to Union and Central Pacific R. R. Cos.	35,000,000	
" to Northern Pacific	47,000,000	
" Atlantic and Pacific	42,000,000	
		124,000,000
" in aid of Canals	4,405,986	
" in aid of Waggon Roads	3,782,213	
		8,188,199
Total		190,296,780
Add grants just made by 41st Congress		33,760,000
Total of all grants to date		224,056,780

The amount received by the different States, made the grantees of these lands, is much less than the figures would indicate. The lands were granted in plots of six alternate sections of 640 acres each, being equal to 3,840 acres to the mile, to be taken by the odd numbers within six miles of the line of the railway. In case a sufficient number of sections of odd numbers of Government lands could not be had, on account of their previous disposal, then the lands of odd sections, within fifteen miles of the railway would be taken, in order to make up the quantity granted. In some cases the grants were enlarged so as to apply to odd sections within twenty miles of the railway. The act of Congress conveying these lands, specified in general terms the route over which the proposed road was to run, and fixed a limit of time for its completion. Owing, therefore, to the condition on which these lands were donated, and the fact that the requisite amount of lands in odd sections within the prescribed limits were not to be had, a number of the companies never received more than half the amount granted them. Of the fifty-eight millions of acres given to the States not one-half has been appropriated as intended, chiefly for the reason just named. The Northern Pacific, which is to run from the head of Lake Superior, through the States and Territories, intervening, to Pugets Sound has the right to take alternate sections within twenty miles of the railway in the States and within forty miles in the Territories, the total grant being 74,423 square miles.

Besides all this liberality on the part of the General Government, the State governments have in many instances contributed handsomely for the encouragement of railway enterprise. We have noticed that the State of Georgia appropriated some thirty millions of dollars in this way, the grants ranging from $8,000 to $15,000 per mile. About two-thirds of this sum was granted at a single session of the Legislature. Alabama guarantees 8 per cent interest on one of her railways, to the amount of $16,000 per mile of completed and equipped railway; another road in the same state has a guarantee covering an expenditure of $20,000 per mile.

PROGRESS OF RAILROADS IN THE UNITED STATES, FROM 1841 to 1869.

STATES.	1841.	1842.	1843.	1844.	1845.	1846.	1847.	1848.	1849.	1850.	1851.	1852.	1853.	1854.	1855.
N. E. States......	589	811	865	865	973	1032	1255	1276	2073	2508	2800	2973	3153	3259	3469
Middle States	1837	2006	2018	3094	2100	2148	2350	2518	2901	3202	3795	4328	4745	5058	5473
Western States......	196	244	280	312	374	419	608	679	727	1276	1846	2426	3708	4001	4567
Southern States ...	913	965	1022	1106	1186	1331	1415	1523	1664	2035	2511	3181	3754	4411	4857
Pacific States	....	......	......	......	......	......	......	....	......	......	.. ...	.... .	.....	.. ...	8
Grand Total ...	3535	4026	4185	4377	4633	4930	5598	5996	7365	9021	10982	12908	15360	16720	18374
Increase of mileage	771	491	159	192	256	297	669	397	1369	1656	1961	1926	2462	1360	1654

PROGRESS OF RAILROADS IN THE UNITED STATES.—Continued.

STATES.	1856.	1857.	1858.	1859.	1860.	1861.	1862.	1863.	1864.	1865.	1866.	1867.	1868.	1869.
N. E. States	3577	3599	3616	3632	3660	3697	3751	3793	3793	3834	3868	3938	4019	4301
Middle States......	5686	6068	6348	6413	6706	6963	7263	7615	7941	8539	9144	9555	9765	10752
Western States	7024	8186	9495	10427	11064	11320	11657	12221	12497	12847	13621	15226	16889	19765
Southern States.....	5707	6627	7386	8274	9182	9283	9422	9468	9511	9632	6867	10102	10683	11272
Pacific States	23	23	23	23	23	23	27	73	166	233	327	431	889	1164
Grand Total	22017	24503	26968	28789	30635	31256	32120	33170	33908	35085	36827	39276	42255	47254
Increase of mileage	3643	2491	2460	1821	1846	621	864	1050	738	1177	1742	2449	2979	4999

RAILWAYS OF THE UNITED KINGDOM FOR 29 YEARS.

Year.	Capital expended on Railways open for traffic.	Average cost per mile.	Total traffic Receipts.	Average receipts per mile for the year.	Working expenses, rates and taxes.	Length of Line open at end of year.	Per centage of traffic receipts to capital expended.	Per centage of net receipts on capital expended.
	£	£	£	£	Per ct.	Miles.	Per ct.	Per ct.
1842	54,380,100	33,362	4,470,700	2743	40	1,680	8·22	4·88
1843	60,687,100	34,925	5,072,650	2995	40	1,780	8·28	4·94
1844	66,852,100	34,290	5,814,940	2982	40	1,950	8·70	5·22
1845	73,646,100	32,736	6,869,270	3080	40	2,243	9·18	5·46
1846	87,763,100	30,903	7,915,870	2797	42	2,840	9·06	5·26
1847	114,722,000	30,924	9,277,670	2501	42	3,710	8·08	4·69
1848	154,200,000	33,223	10,445,100	2258	42	4,626	6·77	4·03
1849	197,000,000	33,110	11,683,800	2000	42	5,960	5·98	3·44
1850	220,522,730	34,236	13,142,235	1944	42	6,738	5·70	3·31
1851	236,841,420	34,166	14,997,310	2162	42	6,928	6·32	3·67
1852	248,093,520	33,816	15,543,610	2118	45	7,337	6·27	3·44
1853	263,[illegible]	33,912	17,920,530	2306	44	7,774	6·[illegible]	3·[illegible]
1854	273,860,000	34,115	20,000,000	2491	46	8,053	7·30	3·93
1855	293,908,000	35,474	21,423,315	2562	47	8,285	7·28	3·86
1856	302,946,200	34,658	23,093,500	2642	48	8,741	7·62	3·96
1857	311,158,670	33,204	24,164,465	2579	48	9,371	7·77	4·04
1858	319,950,000	33,509	23,863,764	2499	48	9,560	7·46	3·88
1859	328,219,100	32,871	25,670,753	2573	48	9,983	7·82	4·07
1860	337,527,300	32,640	27,676,783	2674	47½	10,360	8·19	4·30
1861	352,396,100	32,478	28,563,374	2632	48	10,850	8·16	4·24
1862	370,107,200	32,263	28,950,612	2527	48	11,470	7·83	4·07
1863	387,244,200	32,264	30,798,000	2545	48	12,104	7·95	4·13
1864	402,998,680	32,290	33,502,497	2646	47	12,682	8·30	4·36
1865	433,558,100	32,870	35,626,528	2702	46	13,189	8·22	4·37
1866	463,748,800	34,030	37,513,937	2776	48.8	13,624	8·16	4·17
1867	479,167,200	34,177	39,170,540	2794	50.6	14,020	8·11	4·01
1868	486,998,400	34,222	39,823,266	2800	49.5	14,233	8·18	4·18
1869	494,860,000	34,297	41,505,661	2876	47.5	14,414	8·42	4·42
1870	504,281,000	34,540	43,623,605	2909	48.1	14,610	8·65	4·49

RAILROADS OF THE WORLD.

Statement, giving a list of all countries in which Railroads have been constructed, and showing the length and cost of these works:—

Countries and States.	Length.	Total Cost.	Cost per Mile.
NORTH AMERICA.			
United States of America	47,254	$2,041,225,770	$44,255
Dominion of Canada — Ontario	1,407	107,816,774	75,344
Dominion of Canada — Quebec	575	43,016,519	74,811
Dominion of Canada — N. Brunswick	226	6,954,232	30,771
Dominion of Canada — Nova Scotia	145	6,955,178	47,969
United States of Mexico	202	11,093,840	54,920
WEST INDIA ISLANDS.			
Island of Cuba	431	22,458,548	52,108
Island of Jamaica	14	391,174	27,941
SOUTH AMERICA.			
United States of Columbia	48	8,000,000	166,667
Republic of Venezuela	32	2,758,784	86,212
British Guiana	60	5,539,140	92,319
Empire of Brazil	512	102,992,384	201,157
Republic of Paraguay	46	4,130,340	89,790
Republic of Peru	101	5,697,410	56,410
Republic of Chili	394	24,155,746	61,309
Argentine Republic	231	12,455,058	53,918
EUROPE.			
U. K. G. Britain and Ireland	14,247	2,511,314,435	176,269
French Empire	9,934	1,576,664.892	158,714
Kingdom of Spain	3,429	367,437,924	107,156
Kingdom of Portugal	522	52,887,474	101,317
Swiss Republic	897	78,157,928	87,132
Kingdom of Italy	4,109	382,580,772	93,108
Roman States	216	18,643,472	86,317
Kingdom of Prussia	5,926	747,689,346	126,171
North German States (other)	1,311	117,107,697	89,327
South German States	2,681	234,914,279	87,659
Austrian Empire	4,429	327,369,535	73,915
Kingdom of Belgium	1,703	182,198,861	106,987
" " Holland	881	85,634,081	92,201
" " Sweden	1,194	74,539,032	62,438
" " Norway	44	4,055,656	92,174
" " Denmark	401	22,902,714	57,114
Empire of Russia (in Europe)	8,700	1,448,356,214	166,477
Ottoman Empire (in Europe)	319	14,936,551	46,729
Kingdom of Greece	100	5,000,000	50,000
ASIA.			
Turkey in Asia	143	6,964,243	48,701
Persia	100	6,000,000	60,000
British India	4,092	391,888,791	95,769
Java	102	7,650,000	75,000
Ceylon	37	2,280,530	61,636

RAILROADS OF THE WORLD.—Continued.

Countries and States.	Length.	Total cost.	Cost per Mile.
AFRICA.			
Egypt	[illegible]	$45,168,879	$[illegible],804
Algeria	[illegible]	1,025,024	[illegible]
Cape Colony	85	7,[illegible],792	92,168
Natal	2	119,422	59,711
AUSTRALIA.			
Victoria	400	46,549,200	113,810
New South Wales	174	14,007,522	80,[illegible]
Queensland	102	10,161,610	[illegible]
South Australia	87	5,142,427	59,108
New Zealand	17	1,491,492	87,7[illegible]

RECAPITULATION.

Countries and States.	Length.	Total cost.	Cost per Mile.
North America	49,801	$2,287,061,818	$46,[illegible]
West India Islands	445	722,819,[illegible]	50,248
South America	1,424	165,739,802	116,[illegible]
Europe	61,048	8,252,[illegible],868	185,180
Asia (containing R. R.)	4,474	414,793,[illegible]	92,700
Africa	568	54,937,917	94,[illegible]
Australia	789	77,352,188	98,063
Aggregate in world	118,550	11,455,104,379	96,610

GOVERNMENT EXPENDITURE ON COMMON ROADS IN CANADA.

The following sums were expended on common roads out of the Imperial guaranted sum of £1,500,000, obtained in 1841, and the proceeds of Provincial Debentures afterwards issued. It would be an error to classify all these as Macadamized roads; for some of them on which large sums were expended were, in the greater part of their length, only graded, and provided with necessary bridges across the streams; while others were Macadamised in their entire length. One engineer, Mr. Taggart, describes the process as making the common roads as consisting of putting the branches of trees across the common roadway, with a cover of tender twigs on the top and plenty of earth over all.

	1841. Int. of Imperial Guaranteed Loan Under 4 & 5 Vic., Cap. 28.	1846. Int. of Residue of Imperial Loan and other monies to be raised by Debentures under 9 Vic., c. 63, 64 and 66.
	Sterling.	Currency.
Bay of Chaleurs Road	£15,000 0 0	
Gosford Road	10,000 0 0	
North Toronto Road (Yonge street)	30,000 0 0	£6,500 0 0
Cascades and Coteau du Lac Road	15,000 0 0	594 4 2 52 13 0
Brantford and London Road	55,000 0 0	
London and Port Sarnia Road	15,000 0 0	
London, Chatham, Sandwich, and Amherstburg Road	36,000 0 0	2,181 9 3
Cornwall and L'Orignal Road	1,500 0 0	1,157 3 2
Hamilton and Port Hope Road (including, in 1846, Caledonia Bridge)	03,000 0 0	1,904 6 5 358 5 11
Gaspe Roads		4,564 0 0
Chemin des Caps		500 0 0
Arthabaska Road		10,761 0 0
St. Johns and Stanstead Road and adjacent Roads		9,800 0 0
Grand River Swamp Road		1,000 0 0
Rouge Hill and Bridge		1,500 0 0
Bytown and L'Orignal Road with a bridge over the Rideau		2,939 0 0
Main Eastern Townships Road from Chambly to Granby		24,889 0 0
Dover Road		325 19 9
Toronto and Saugeen Road		71 6 6
Rice Lake and Ontario Road		123 17 1
Port Stanley Road (Toll Houses)		50 0 0
Rondeau Road		1,969 1 2
Bridges between Quebec and Montreal	34,000 0 0	
Bridges South of St. Lawrence (over River Etchremin, Nicolet Becancour, Godsfroy, Chateauguay, and Duchene		7,300 0 0
Bridge over River Champlain		500 0 0
Jacques Cartier Bridge		1,000 0 0
Lancaster Bridge		170 0 0
Bayonne Bridge		144 4 10
Gananoque Bridge		7 3 0
Chaudiere Bridge		307 9 0
Union Suspension Bridge, at Bytown		91 7 7

CANADIAN CANALS.

STATEMENT showing the Tonnage and the Tolls levied on Freight and Passengers passed through all the Canadian Canals from the year 1850 to 1870, inclusive, distinguishing whether from or to Canadian or United States Ports; also the Tonnage passed free.

Remarks.	Years.	From Canadian to Canadian Ports. Tons.	From Canadian to American Ports. Tons.	From American to Canadian Ports. Tons.	From American to American Ports. Tons.	Net Tonnage and Tolls Collected on Freight. Tons.	Net Tonnage and Tolls Collected on Freight. Tolls. $ cts.	Tolls Collected on Freight and Vessels. $ cts.	Net Revenue on Freight and Vessels. $ cts.
Ottawa Canals not included under control of Imperial Gov'nment.	1850	538,477	224,335	52,183	221,895	1,037,390	239,898 16	236,123 58	238,138 58
	1851	630,212	217,500	103,962	265,130	1,416,794	279,229 70	304,864 90	304,864 90
	1852	896,030	153,006	34,858	449,720	1,497,614	309,316 64	343,306 96	330,409 32
	1853	1,016,902	241,901	55,211	403,496	1,865,709	347,622 78	390,487 02	382,915 80
	1854	1,006,906	149,636	118,663	412,999	1,687,304	292,966 39	333,101 85	332,081 87
	1855	849,967	303,923	101,361	437,623	1,672,404	301,153 77	326,090 21	304,081 48
	1856	922,696	250,741	230,343	542,812	1,904,572	342,930 47	391,250 59	361,382 66
	1857	856,098	264,141	215,566	423,825	1,759,625	298,718 97	305,091 28	330,167 83
Season of Navigation.	1858	1,424,313	299,587	162,984	471,432	2,259,709	262,861 36	314,447 51	263,410 50
	1859	1,663 599	336,487	126,707	380,601	2,409,342	190,657 15	236,962 41	275,427 26
	1860	1,331,186	473,365	169,671	589,479	2,363,701	159,098 63	338,398 56	296,921 96
	1861	1,522,079	246,742	299,796	546,393	2,614,892	136,615 31	419,386 02	155,381 75
	1862	1,545,219	552,606	371,510	644,393	3,113,728	154,565 55	407,305 96	213,487 41
	1863	1,644,908	491,112	274,597	621,366	3,032,275	337,607 19	365,220 31	365,328 21
½ year.	1864	490,449	137,756	69,112	158,704	812,496	86,326 64	106,611 80	103,976 75
Fiscal years.	1865	1,090,930	483,575	346,463	367,840	2,428,814	241,346 66	290,966 47	262,757 17
	1866	1,320,111	671,042	194,404	465,713	2,659,272	284,554 70	318,867 74	304,268 24
	1867	1,690,816	706,067	284,293	481,074	3,127,670	276,681 13	325,263 42	316,420 54
	1868	1,788,428	810,939	279,706	644,946	3,467,116	316,821 73	331,129 15	306,778 28
	1869	1,716,829	712,946	368,221	689,061	3,484,877	318,348 06	309,089 10	367,806 32
	1870	2,088,571	888,878	330,794	686,350	4,015,698	361,710 58	444,983 26	607,480 28

CANADIAN CANALS.

STATEMENT shewing the Tonnage of, and Tolls levied on Vessels passed through all the Canadian Canals, from the year 1850 to 1870, inclusive; distinguishing whether from or to Canadian or United States Ports.

Remarks.	Years.	From Canadian to Canadian Ports.	From Canadian to American Ports.	From American to Canadian Ports.	From American to American Ports.	Tonnage.	Tolls Collected on Vessels.
							$ cts.
Ottawa Canals not included under control of Imperial Gov'nment.	1850	592,865	179,242	173,491	244,873	1,194,475	18,225 42
	1851	1,167,958	259,577	175,240	371,066	1,973,841	25,635 20
	1852	1,274,181	305,521	238,661	468,628	2,286,991	29,192 68
	1853	1,123,804	284,108	168,434	562,308	2,138,654	35,292 72
	1854	1,287,039	221,229	118,510	475,845	2,102,623	39,815 18
	1854	1,387,747	219,710	271,389	479,168	2,358,014	33,537 65
	1856	1,240,521	300,625	495,371	636,542	2,673,086	38,661 61
	1857	1,132,400	327,950	341,818	513,726	2,315,894	36,368 36
Season of Navigation.	1858	1,444,578	306,294	307,270	654,224	2,712,366	39,549 23
	1859	1,474,385	344,604	223,241	412,791	2,455,021	32,740 23
	1860	1,586,767	430,198	284,950	728,815	3,030,730	46,828 30
	1861	1,954,193	322,214	317,061	714,443	3,307,911	49,703 44
	1862	2,015,342	391,663	387,475	787,916	3,582,396	58,603 86
	1863	1,948,803	444,864	471,445	708,801	3,573,913	47,523 02
½ year.	1864	675,982	123,460	104,234	212,874	1,116,550	15,240 11
Fiscal years.	1865	2,231,304	369,031	350,630	469,810	3,420,835	41,411 51
	1866	2,195,158	530,215	383,495	483,161	3,592,029	39,797 54
	1867	2,414,910	536,251	415,160	477,451	3,843,772	41,788 41
	1868	2,514,446	631,094	406,178	634,292	4,186,010	48,954 59
	1869	2,430,705	652,080	474,931	649,992	4,207,708	49,208 14
	1870	2,730,614	785,581	535,434	690,493	4,752,522	55,752 64

EARLY RAILWAY CHARTERS.

The first railway legislation in Canada commenced with the incorporation of *The Company of Proprietors of the Champlain and St Lawrence Railway* (2nd Wm. IV, cap. 58) on the 25th February, 1832. The capital was £50,000, in £50 shares, with power to increase to £65,000 if necessary. The line was to commence at or near the village of St. Johns, and run to or near the village of Laprairie, or to some point between it and the upper end of St. Helen's Island. The time for the work to be commenced was at first fixed at three years, and afterwards at four. After the expiration of one year, after the opening of the road, the tariff for the carriage of goods and passengers was to be yearly regulated by the amount of dividend declared in the preceding year. This original charter underwent several successive amendments. [By 13 and 14 Vic., Cap. 13, the privileges granted to the Montreal and Province Line Railway Company, was (July 24, 1850) transferred to the Champlain and St. Lawrence Railroad Company; the latter being authorized to construct a branch from some point of their line to the River St. Lawrence, opposite Montreal; and to continue their railway from at or near the terminus at St. Johns to Rouse's Point, there to connect with an American Railway then in course of construction. This would give an uninterrupted line of communication between Montreal and Boston, New York and the West. The second Railroad Company chartered in Lower Canada, was the Quebec and Province Line Company. The line was to run into the State of Maine, near Monument Stream, and to be completed in five years. The Act expired by non user.]

The Canada Union Railroad Company was chartered by ordinance of Special Council of Lower Canada, June 26, 1840, with a capital of £100,000 in £25 shares, and power to raise a further sum of £25,000 if necessary. [The road was to run from Montreal to the division line between Upper and Lower Canada, at or near Point-a-Beaudet, though it might terminate at Coteau-du-Lac should the Company think that length would give sufficient facilities for transport to Upper Canada. The time for building the road was afterwards extended to 1847.]

The Company of Proprietors of the Upper Ottawa Railroad Company was chartered by ordinance of Special Council of Lower Canada, June 26, 1840, with a capital of £30,000 in £50 shares, and authority to raise £15,000 more if required. The line was to run from or near Carillon, on the Ottawa, to or near Grenville. Act expired by non-user.

The Carillon and Grenville Railroad Company was chartered on the 24th June, 1846, with a capital of £60,000 in £25 shares. The line was to run from some place in the County of Two Mountains, from Carillon to Grenville. This charter contained a singular clause, which shows that the promoters or the Government, if not both, had a very erroneous idea of the profitableness of railroads. Whenever the stockholders had got ten per cent. dividend, the Company was thenceforth to pay over to the Government a moiety of their net earnings exceeding three pounds currency a share, as a tax.

The Company of Proprietors of the Eastern Townships Railroad was incorporated by ordinance of special council, Jan'y. 21, 1841, with a capital of £150 000 in shares of £12.10s. each, and authority to raise £40,000 more if necessary. The road was to run from the town of Sherbrooke, by way of the outlet of Lake Memphremagog, in the county of Stanstead, to some point on the River Richelieu. The Company might build bridges over the Magog and Richelieu, but not obstruct the navigation of the latter, or interfere with the privilege granted to the Champlain and St. Lawrence Railroad Co.

The St. Lawrence and Atlantic Railroad Co. was incorporated (8 Vic. Cap. 25,) on the 17th March, 1845, with a capital of £600,000, in £50 shares, and was empowered to raise a further sum of £500,000 if necessary. By a subsequent Act, the shares were reduced to £25 each. The line was to run from the River St. Lawrence, as nearly opposite Montreal as may be found desirable, in the general direction of St. Hyacinthe, to such point on the frontier line as would admit of the most advantageous connection with the *Atlantic and St. Lawrence Railroad*. They were also authorized to construct a branch line to the boundary line in the county of Stanstead, with a view to a connection with any railroad that might be constructed in the State of Vermont; and a second branch from the south bank of the St. Lawrence opposite the city of Quebec. As in the case of the first railroad company chartered in Canada, the passenger and freight rates were to be regulated by the amount of the dividends. Even then the idea of a railroad bridge across the St. Lawrence was floating in men's minds; and this Company was authorized, in the event of such bridge, to construct a branch road to the south end of it; and from the end of the bridge on the other side of the river to the city of Montreal, and to agree with the Bridge Company for liberty to use the bridge. The corporation of Montreal or ecclesiastics of the Seminary of St. Sulpice, or any other corporate bodies civil or ecclesiastical, were empowered to take stock or lend money to the Company the first instance of such authority being given in Canada. It was not till some years after that a like general authority was given to municipal corporations in Upper Canada.

The Montreal and Lachine Railroad Company was chartered (9 Vic. cap. 82) on the 9th June. 1846, with a capital of £75,000, in £50 shares, and power to increase to £100,000 if necessary. When the £75,000 had been expended, the Company's stock had depreciated, so that £40,000 instead of £25,000 more capital was found necessary, and was authorized to be raised by 12 Vic. cap. 177. The title of the Company sufficiently indicates the location of the road. There was a clause in the charter taxing the revenues of the Company for the benefit of the Government out of earnings which might exceed ten per cent. on the capital; a delusion similar to that contained in the St. Lawrence and Atlantic Company's charter. Authority was given to transfer the property of this Company to another company, which should have expended not less than £100,000 in the construction of a railroad from Lachine viâ Prescott to Kingston; but such sale was not to operate as a dissolution of the Company—the purchasers were to become the Company. All corporations, civil or ecclesiastical, were authorised to subscribe for the new (£40,000) stock. By another Act (13 and 14 Vic. cap. 12) the Company was authorised to amalgamate with the Lake St. Louis and Province Line Railroad Company, and to a new company to be called The Montreal and New York Railroad Company. By 13 and 14 Vic. cap. 113, the Montreal and Lachine Company was authorised to extend its line of railway to or near Prescott, and to possess one steam or other vessel to ply on the rivers Ottawa and St. Lawrence, or either of them, in connection with the road. For these purposes the Company was authorised to issue stock to the amount of £750,000 in shares of £12 10s., in addition to the stock previously authorised. As soon as they had completed twenty-five miles of the additional road, the Company was to change its name to *The St. Lawrence and Ottawa Grand Junction Railway*; but it was not thereby to become a new corporation. But if this Company could not undertake the proposed extension, then certain persons, who were named, were to be incorporated for that purpose, under the name of *The St. Lawrence and Ottawa Grand Junction Railroad Company*.

The Lake St. Louis and Province Line Railway was chartered (10 and 11 Cap. 120) June 24, 1848, with a capital of £150,000 in £50 shares, with Vic. power to raise £30,000 more if necessary. The line was to run from the village of Sault St. Louis, in the County of Huntington, to some convenient point in the counties of Huntington or Beauharnois, within three miles of the line dividing the township of Hemingford from the County of Huntington, with a view to its junction with some railway to be constructed to connect the North-Western part of the State of New York with Lake Champlain. The Company was to pay as a tax to the Government a moiety of its net income exceeding six pounds a share, as soon as the dividends, on the whole, should have amounted to ten per cent. on the stock. By the 13 and 14 Vic. Cap 112, this company was authorised to amalgamate with the Montreal and Lachine Railroad, on conditions stated above.

The St. Lawrence and Industry Village Railroad was incorporated (10 and 11 Vic. Cap. 64) on the 28th July, 1847, with a capital of £12,000 in £25 shares, and power to raise a further sum of £4,000 if required. It was to run from some place on the river St. Lawrence, in the parishes of Lavaltrie or Lanoraie, district of Montreal, to some place in the parish of St. Charles Borromée, at or near Industry Village. As soon as the dividends had amounted, in the whole, to ten per cent. on the paid up stock, a moiety of the net income, exceeding six pounds a share was to go as a tax to the Government.

If Upper Canada was behind the sister Province in railroad legislation, the distance was not great.

The Cobourg Railroad Company, whose line was to run from any point on Rice Lake to Lake Ontario, at or near Cobourg, was incorporated (4th Wm. IV., cap. 28) March 6, 1834,—some twenty years before the line was built,—with a capital of £40,000 in £10 shares. The work was to be commenced within two years and completed in eight. The time for commencing was afterwards extended to April, 1839. By 7 Wm. IV., cap. 74, Government debentures to the amount of £10,000 were authorized to be issued to the Company by way of loan— the first instance in Canada of proffered Government aid in the construction of a railroad. This Act having expired by by *non-user* the Company was revived by 9 Vic., c. 80, with more modest pretensions, under the title of *The Cobourg and Rice Lake Plank Road and Ferry Company.* The object was to substitute a plank or macadamized road for a railway. The capital was £6,000, with power to double that sum if necessary, and unlimited time to perform the work in.

The Toronto and Lake Huron Railroad Company was incorporated (6 Wm. IV. cap. 5) April 20, 1836, with a capital of £500,000 in £12 10s. shares. The railway was to run from the city of Toronto to some portion of the navigable waters of Lake Huron within the Home District. The road was to be commenced in three years and completed in ten. The Government was authorized to issue debentures in aid of this work to the amount of £100,000, during the progress of its construction. If the Company could not meet the interest on the debentures, the amount was to be raised by assessment on Toronto and the country through which the railroad passed. Authority was given to construct a branch to Lake Simcoe. By 7 Wm. IV. cap. 63, this road was divided into three sections. 1. South of the Oak Ridges; 2. North of the same; 3. To the township of Nottawasaga, on Lake Huron, or to the terminus in the county of Simcoe. No one of these sections was to be commenced until the preceding had been completed. All the stock subscribed was to be called in within five years. The act of incorporation expired by *non-user.* But here, as in the case of the Cobourg road, the Company was revived, with a similar lowering of pretensions, which implied that they had been in advance of the times. By 8 Vic. cap. 83, the Company was empowered to construct a Rail, Planked, Macadamized or Blocked road. The directors might make the terminus at any point on Lake Huron. The time for completing the work was extended to March, 1850. The capital of the Company was not reduced, but remained at £500,000. By 10 Vic. cap. 111, the Company was author- to construct one or more branches from the main road, extending westward from Toronto to Lake Huron, giving the road two termini on the western frontier, but both of them were to be north of Sarnia. To enable the Company to carry out this formidable undertaking, the authorized capital was increased to £1,500,000. The liberal allowance of twenty years was given to construct the branches. A declaratory Act (10 and 11 Vic., cap. 66) enacted that parties who had become stockholders in the railroad contemplated by the original Act were not to be considered as subscribers to the stock afterwards authorized to be employed, at the option of the directors, in a different kind of road, though the liabilities incurred under the first Act remained in force. This charter expired by *non-user.*

The Huron and Ontario Railroad Company (6 Wm. IV., c. 7,) was chartered on the 20th April, 1836, with a capital of £350,000, and privilege to extend to £500,000. It was to run from Wellington Square, Burlington Bay, or Dundas, to Goderich. The Company was authorized to construct harbours, moles, piers, wharves, etc., at the termini.

The Niagara and Detroit Rivers Railroad Company was incorporated (6th Wm. IV., c. 6,) April 20, 1836, with a capital of £800,000, in shares of £6. 5s. each. It was to run from the Niagara River, in the township of Bertie, to the Detroit River, at Sandwich. The Hamilton and Port Dover Railway Company, or any other incorporated Company, was authorized to establish lateral branches from its line of railway to Queenston, Niagara, London, Chatham, or any other place between the townships of Sandwich and Bertie. This charter expired by non-user.

The London and Davenport Railroad and Harbour Company was chartered (6 Wm. IV, c. 52,) March 4, 1837, with a capital of £50,000, in shares of £6. 5s. each. It was to run from the town of London to Lake Erie, at or near the village of Davenport, at the mouth of Cat Fish Creek, where it might construct a harbour, moles, piers, wharves, etc. Expired by non-user.

Montreal and Kingston Railroad Company was chartered on the 28th Dec., 1846, (Vic. 10, cap. 107,) with a capital of a million currency ($4,000,000), in shares of £25, or $100, each. The Directors were empowered to make arrangements for uniting with any other Railway Company, then chartered, or to be thereafter chartered, for building a road between Kingston and Montreal, or any part of the distance. If they acquired the property of the Lachine, or any other Railroad Company, their capital was to be increased to the extent of the capital stock thus acquired. This road might be constructed in connection with a contemplated chain of railway from Montreal to the Western boundary of the Province. The road was to be commenced within four years. The charter expired from non-user.

The Wolfe Island, Kingston and Toronto Railway Company (Vic. 10, cap. 108,) was chartered at the same time as the above, with a like capital, similarly divided. Beside connecting Kingston and Toronto, this road was to have a branch from Kingston (by rail and steam ferry) across Wolfe Island, to the boundary line of the Province, opposite Cape Vincent. It might form part of a general line between Montreal and the Western boundary of the Province. The same time was given to commence work as in the previous charter, with the same result.

The Peterboro' and Port Hope Railway Company was chartered (Vic. 10, c. 109) Dec. 28, 1846, with a capital of £100,000, in £10 shares. It was to start from Peterboro' and strike Lake Ontario at or near Port Hope. The road might be extended from Peterboro' to Chemung Lake, in the Colbourne district, any time within ten years.

The Hamilton and Toronto Railway Company was chartered (10 Vic., cap. 118) Dec. 28, 1846, with a capital of £225,000 in £25 shares. The title sufficiently indicates the route. It was authorized to be constructed as a continuation of the Great Western, in view of extending a grand chain of railway communication from the western extremity of the Province to Montreal. Expired from non-user.

The Erie and Ontario Railroad Company was chartered (5 Wm. IV., c. 19) April 16, 1835, with a capital of £75,000, and power to increase it to £180,000 in case of the extension of the line to Erie and Ontario. It was to run from some point on the River Welland to the Niagara River at or below Queenston, with power to extend to Lake Erie or to the Niagara River below Lake Erie, and from Queenston to Lake Ontario, provided they let a contract for the work before any other company was incorporated to perform it. The Legislature expressly reserved the right of incorporating hereafter any other company for constructing a railroad on any other line between Lakes Erie and Ontario, provided it would not impede the progress of any line begun by this Company within three years. Authority was given to the Government by 7 Wm. IV., cap. 68, to loan £5,000 by the issue of debentures to enable the Company to complete the road.

The Woodstock and Lake Erie Railway and Harbour Company was chartered (10 and 11 Vic., c. 117) June 24, 1848, with a capital of £250,000 in £5 shares. The road was to start from the town of Woodstock and run to some point of Lake Erie between the harbours of Port Dover and Port Burwell. The Company was empowered to build steamboats or other vessels for conveying their passengers and freight from any ports on Lake Erie to any other place, and to construct harbours, wharves and piers for the use of such vessels. There was a special clause prohibiting travelling on this road on Sunday.

The Bytown and Britannia Railway Company was incorporated (10 and 11 Vic., c. 118) June 24, 1848, with a capital of £10,000 in £5 shares, with power, if necessary, to raise a further sum of £10,000. The road was to run from Bytown (Ottawa) to some place or places in the township of Nepean, at or near Britannia Mills. The Company was empowered to possess steamboats and other vessels to ply on the Ottawa from the upper terminus of the road, or any place above it, to Fitzroy harbor, and thence to Portage-du-Fort, in the township of Ross. Charter expired by *non-user*.

The Bytown and Prescott Railway Company was chartered (13 and 14 Vic., c. 132,) August 10, 1850, with a capital of £150,000, in £10 shares, with authority to increase to £250,000, if necessary. The road was to run from some place or places on the Ottawa River, at or near Bytown, to some place or places on the St. Lawrence River, at or near Prescott. The Company was authorized to own and run steamers on the Ottawa and St. Lawrence, to run not more than twelve miles from either terminus.

If anything further were wanting to show that the Canadian people were thoroughly imbued with the spirit of enterprise in those days, it might be found in a petition addressed to the Legislative Assembly of Canada in 1854. The petitioners, who were Canadians and Americans, asked incorporation under the name of "The Northern Pacific Railway Company;" and sought power to construct a railway from Montreal, up the Valley of the Ottawa, and along the north shore of Lake Huron, to the Western boundary of Canada. The south shore of Superior was then to be traversed, after which the only indication of route was from the head of Lake Superior and Puget's Sound to the mouth of the Columbia River.

THE RAILWAYS OF CANADA.

1870–71.

*GRAND TRUNK RAILWAY.

CANADIAN RAILWAY LEGISLATION.

[The first impulse given to railroad enterprise was derived from the Railroad Guarantee Act of 1849. That Act laid down a policy on which the Government should assist private companies undertaking the construction of railroads. This Act (12 Vic. Cap. 29) was entitled "An Act to provide for affording the guarantee of the Province to the Bonds of Railway Companies on certain conditions, and for rendering assistance in the construction of the Halifax and Quebec Railway." The preamble recited that, in new and thinly peopled countries, where capital is scarce, Government assistance in the construction of railroads is necessary, and may be safely afforded to lines of considerable extent, in a form of a guarantee to private companies acting under charter. The enacting clauses provided that such aid should not be given towards constructing any railroad less than seventy miles in length. The Province was not to issue debentures or provide capital, in any shape, but merely to guarantee the interest of loans which the railway companies might raise on their own securities; in other words, the Government was to endorse the securities of the companies, but only for the payment of interest, and to the amount of six per cent. The amount of the guarantee was to be limited by the cost of the road; it was not to exceed one-half of the entire cost, and was not to be given until one-half of the road had been completed, and when the amount to be given would be sufficient to complete the road to the satisfaction of the Commissioner of Public Works.

This guarantee Act seems to have been passed on a fallacious hope that the Government would not be put to any great charge in the matter; for security, which turned out to be no security at all, was taken on the revenue of the Company, out of which source a Sinking Fund was to be raised to pay off the debt. The payment of the interest on the guaranteed bonds was to form a first charge on the revenue of the Company, and no dividend was to be

* A tabulated statement showing the traffic, earnings, mileage, &c. &c., of all the Canadian Railways, will be found at the end of these reports.

declared till after the interest was paid, and three per cent. of the capital set apart for a Sinking Fund. The Province was to have a first lien on the road for any sum paid or guaranteed.

Provision was also made respecting the Quebec and Halifax Railway. It was confined to an offer of an annual payment of £20,000 stg., a year towards making good any deficiency in the income of the road, in meeting the interest on the expenditure, and a grant of all lands, in the possession of the Government, along the line of Railway, within the Province, to the extent of ten miles on each side, and to find the land for the right of way and stations within Canada. This offer was conditional on the Imperial Government undertaking, either directly or through the medium of a Company, the construction of the railway. There was no stipulation for any share of Government control as the price of this aid.

An Intercolonial Railway, connecting Halifax and Quebec, was not likely to be built without the Lower Provinces bearing some part in the work. Accordingly, in 1850, the Nova Scotia Government took up the question, and Mr. Howe went to England to try "to obtain the necessary funds from London capitalists, either with or without the aid of Her Majesty's Government," in the words of Sir John Harvey, then Lieutenant Governor of that Province. Mr. Howe, in a letter to Earl Grey, Nov. 25, 1850, expressed the singular opinion that, "if our (Nova Scotia) Government had means sufficient to build railroads and carry the people free, we believe that this would be a sound policy." And he added: "If tolls must be charged, we know that these will be more moderate and fair if Government regulate them by the cost of construction and management, than if monopolies are created and speculators regulate the tolls only with reference to the dividends. We are all wise after the fact, but the extravagance of this opinion will now strike every one. If tolls were high enough to cover the cost—that is pay interest on all the capital expended—there would be little travelling on Dominion railroads. Mr. Howe asked the guarantee of the Imperial Government for the capital necessary to build the Intercolonial, which he put at £5,000,000 sterling. That would produce low tolls, of which every Englishman travelling over the line would get the benefit. If this aid was rendered, the Queen's name would become a tower of strength on the continent, but if we had to borrow largely from America, a revulsion of feeling dangerous to British interests would be created. To refuse would wound the pride of every Nova Scotian, and cause other mischiefs. The Canadian rebellion had cost £5,000,000 to put down; and if the Maritime Provinces had joined in the fray, they could have cut off every regiment that marched through them in 1837 and 1839. Nova Scotia had offered her entire resources in aid of Imperial policy when, in 1849, peace was rendered insecure by the state of the Maine and New Brunswick boundary question. Were these people now to be subjected to foreign capital and control? Americans were willing to do the work, and take stock in payment. They would embark in it for the sake of national control. And if they did so they would import republican ideas, which no Nova Scotian or New Brunswicker would deem it worth his while to counteract. Annexation would destroy the cursed Yankee monopoly, which had pressed so heavily on Nova Scotia, in an hour; and the Nova Scotia fisherman would be secured in a bounty that would cover his risk, besides a free market without bounds. Construct the road, and the drooping spirit of the colonists would revive.

At the distance of twenty years, it seems almost incredible that any colonist should have gone to England with such arguments in his mouth. But Mr. Howe thought this was arguing the case on its merits. Another letter followed (Jan. 16, 1851) from the same pen, with more hints about annexation, and a vast quantity of matter, mainly irrelevant.

Mr. Hawes, Under-Secretary of State for the Colonies, replied, March 10, 1851, stating the conditions on which the Imperial Government would guarantee an Intercolonial Railway loan. Nothing would be done until arrangements were made with Canada and New Brunswick for extending the road from Halifax to Quebec. The Government would recommend to Parliament the giving of like assistance to all the three Provinces. The line was not necessarily to be that of Major Robinson, but any deviations from it were to be subject to the approval of Her Majesty's Government. There would be no objection to this road connecting with one to the United States. The Provinces were to make the guaranteed loans; they would require to raise a first charge on their respective revenues, after existing obligations, and provide a Sinking Fund for their extinction. The money was to be spent under Commissioners to be appointed by the guaranteeing authority. Troops, stores and mails to be sent over the road at reasonable rates.

Four days after, Earl Grey, then Secretary of State for the colonies, addressed a despatch to Lord Elgin, Governor-General of Canada, in which he stated: "it is necessary to ascertain whether Canada and New Brunswick are ready to join with Nova Scotia, in raising the capital required for the work in the manner proposed, and if so in what proportion each Province is to become responsible for the expense incurred."

The conditions on which the Imperial Government was prepared to give the guarantee, left the local Governments to proceed on the basis of the Canadian Guarantee Act of 1849 and construct the railroad through the instrumentality of a company or companies, if they saw fit. But a change was now determined on in Canada. In 1849, it had been declared in the preamble of the Guarantee Act, that Government assistance was "best given by extending to companies engaged in constructing railways" the guarantee already described. Now the policy changed, and Mr. Hincks declared that "the experience of other countries warranted the conclusion that the best mode of constructing and managing railroads was by placing them under the control of the State;" an opinion which would now, and perhaps then, gain the consent of but very few statesmen in other countries. Canada, he assumed, would have the control of four millions sterling as her share—Mr. Howe assumed the whole would be £7,000,000—of guaranteed capital with which to build her share of the Intercolonial railroad; and that any thing she could save out of this amount would be at her disposal to aid in building other railroads; and he assumed that the amount would be sufficient to extend the Intercolonial west ward as far as Toronto. The estimates on which he relied were extremely low; less than one-half the actual cost. From Toronto to Kingston, Mr. Keefer's estimate was £4,500 a mile; from Kingston to Montreal £5,000 a mile. But Mr. Hincks, to be on the safe side, estimated both sections at £5,000; and for the section from Melbourne to Quebec he put down £6,000 a mile. From Quebec to Halifax, he assumed the cost at £7,000 a mile. These items made a total of only £3,333,000; to which, by way of extra precaution, he added enough to bring the whole up to the round figure of £4,000,000.

There was now, 1851, passed (14 and 15 Vic., Cap. 73), *An Act to make provision for the construction of a Main Trunk Line of Railway throughout the whole length of this Province.* This act brought the Legislature under a pledge not to increase the public debt, except for the purposes of building such railway, and "as regards the guarantee of the Province under the Act 12 Vic., Cap. 29, for interest only on debentures issued or to be issued by the *St. Lawrence and Atlantic, the Great Western, or the Ontario, Simcoe and Huron Railway Companies.*" The Governor General was authorized to enter into arrangements with the Governments of Great Britain, and of the Lower Provinces, for the construction of the Quebec and Halifax Railway, if the necessary funds should be raised under the Imperial guarantee. The Governor in Council was authorized to apply, in furtherance of that work, all the ungranted lands, to the extent of ten miles on either side of the line. The road was to be continued as far as Hamilton, under the Imperial guarantee, if that were obtained; but if it was not obtained, or the amount was not sufficient to accomplish so much, the whole road, or the residue of it, was to be built at the joint expense of the Province, and such Municipal Corporations as would subscribe towards it. A fund was to be formed out of the municipal subscriptions, to be called the "Municipal Subscription Fund." Debentures equal in amount to these municipal subscriptions might be issued by the Government, and chargeable on this fund, and a Sinking Fund to be created; besides an equal amount of debentures chargeable on the consolidated revenue. If the funds for constructing the Main Trunk could not be raised in any of these modes, the work might be undertaken by chartered companies. A Board of Railway Commissioners, consisting of the Receiver General, the Inspector General, the Commissioner and the Assistant Commissioner of Public Works, was created. The guarantee under the Act of 1849, was not to be given till this Board had reported to the Governor in Council, that the land for the whole line or section had been obtained and paid for, and a part of the work done; and that the fair cost of this was equal to what would have to be expended for the completion of the road.

The Government had set out, as we have seen, in 1849, by confining the guarantee to the interest of the loan raised by the railway company; but by the Act of 1851, now under review, authorized the Governor in Council to extend it to the principal, in case of the Grand Trunk. Provincial debentures might be exchanged for those of railway companies. In return, the Province was to take the delusive security of a first lien on the railway, tolls and property of the Company ; a security from which the Province has never derived and never will derive a single dollar. We now know that the straightforward way of dealing would have been to grant a bonus instead of a loan that purported to be secured. Whatever the Province has paid on account of this road it has got good value for ; but the mode of doing it held out hopes that have not been realized.

QUESTION OF ROUTE.

The question of the route of the Main Trunk engaged the attention of the Standing Committee of the Canadian Legislature on Railroads and Telegraph Lines in 1851. There was much contrariety of opinion as to where the section of the line between Kingston and Montreal should be located. Mr. Hugh Allan thought it should be so located that it would serve the purpose of opening up and accommodating the Ottawa District. He thought the northern route by Bytown (Ottawa) would secure the largest amount of traffic, as it was further removed from the competing water communication of the St. Lawrence. The cost would be more, but he thought the increased traffic would ultimately compensate for the difference. Large tracts of land would be opened for settlement, and easy access to market for the produce given. For through business he was obliged to admit a line running parallel with the St. Lawrence would be the best ; but if that were constructed subsidiary lines should be constructed to connect with Bytown, Perth and other places. Several other witnesses spoke to the same effect. One argued that if the northern route were not adopted, a second, interior main line, would ultimately be constructed. The military argument was also resorted to, and much was said of the security of an interior and the insecurity of a frontier line. On the other side it was contended that the true interests of the Province would be best consulted by adopting the most direct route, because experience showed that a contrary course was prejudicial to through traffic. Very little business would take a northern direction. Everything would go south ; and the cost of construction, mile for mile, independent of the increased length of 20 or 25 miles, would be greater on the northern route, and no trunk line could accommodate all the traffic along the way without the aid of auxiliary lines. The question has not now sufficient interest to warrant us in recapitulating all the arguments in favour of the interior route : it is sufficient to say that opposed to them was the single argument that the short line was best for the direct trade ; and this outweighed everything else in the selection of the route.

Mr. Gzowski, who favored the St Lawrence line—that actually adopted—estimated the cost of construction at £5,025—not quite $20,000—a mile ; and there were people who entered into elaborate arguments to prove that this was too much. Mr. T. C. Keefer's estimate was £5,340 a mile, including "ample equipment." For the Kingston and Toronto section his estimate was still lower, but it was a different class of road from that finally built. But the Atlantic and St. Lawrence road had cost £7,000 currency—$28,000—a mile, and it was estimated there would have to be a further expenditure of about $5,000 a mile.

QUESTION OF GAUGE.

On the question of gauge, several witnesses were heard. We incline to think that the weight of the evidence was in favour of a four feet eight and a half inch gauge, while that of five feet six was adopted. Even Mr. T. C. Keefer did not venture to suggest a greater breadth than five feet, while expressing the opinion that time would vindicate the sufficiency of the narrow gauge, and most of the authorities to which he referred, including that of Robert Stevenson, were in favor of the narrow gauge. Mr. Keefer himself said: "The steadiness of a carriage depends upon the length of the rectangle formed by the wheels, and I think the long carriage used on the American narrow-gauge roads are steadier than the short broad gauge carriages, when both are run upon roads of equal condition." A Royal commission, appointed in 1845—six years before—had reported: "that as regards the safety, accommodation and convenience of passengers, no decided preference was due to either gauge; that in respect to speed, the advantage was with the broad gauge; that in the commercial case of the transport of goods, we believe the narrow gauge to possess the greater convenience, and to be more suited to the general traffic of the country; that the broad gauge is the more costly;" and they ended with this conclusion: "Therefore, estimating the importance of the highest speed on express trains for a comparatively small number of persons—however desirable it may be to them—it is of far less moment than affording increased convenience to the general traffic of the community—we are inclined to regard the narrow gauge as that which should be preferred for the general convenience." The question was here one between a 4 feet 8½ inch and a 7 feet gauge, and the commissioners took care not to express an opinion in favor of the 4 feet 8½ inch gauge in preference to all intermediate or possible gauges.

Many of the persons examined before the Assembly committee, in 1851, were not in a position to form the best opinion as to the relative value of different gauges. Mr. Harris, President of the Great Western, must be presumed to have given the question some consideratiou, and he gave his opinion in favour of the narrow guage, which the Great Western had then adopted. All their calculations, plans and specifications were then based on a four feet eight and a half-inch track. He gave the following as the reasons for its adoption:

"First, its established character; second, the saving of money in the superstructure (ties and rails requiring extra strength for broader gauge); third, saving of expense in running machinery, for all time to come; and fourth, to form an easy and economical junction with the railroads of Michigan and New York, from which the Company expect to receive very large additions to the traffic on their road, a considerable portion of which is expected to follow a Trunk Line through the Province to Montreal." And he added:

"I consider the adoption of a broader gauge than four feet eight and a half inches would prove injurious to the interests of the Great Western Company, as well as to the Main Trunk Line as far as Montreal, because I feel that every inducement possible will require to be made, to securethe principal part of the travel from Chicago, &c., through Canada, in preference to the various channels now being opened on the south side of Lake Erie; and I feel convinced that any gauge that will not admit of the luggage cars of the roads joining the Great Western on either side being carried across it, will deprive Canada of the greater part of the said travel.

"I think a uniform gauge from Windsor to Montreal very important, as ensuring to 'through' American travel the expedition so much prized at the present time; and if this gauge afforded an easy and economical junction at Detroit, I feel confident a very large and remunerative passenger trade would be established, highly beneficial, in every way, to the Province, part of which would diverge at Hamilton, part at Toronto, part at Kingston, &c., and still a large portion would go as far as Montreal, but no through (American) passenger trade of consequence would go beyond the latter point. This trade can only be got, however, by amicable and mutually beneficial arrangements between the Railroad Companies in the United States and the Companies that join them on the Canada side; I do not, therefore, consider it of much consequence whether the same gauge is continued on the south side of the Saint Lawrence between Montreal and Quebec, or not, and more particularly as the importance of the City of Montreal would prevent any number of passengers, either on business or pleasure, passing the said city, without stopping a longer time than could be allowed by a junction train."

There is something prophetic in some of these reasons. The Great Western, practically compelled by the Legislature to adopt a five feet six gauge, were obliged to reduce it, by means of a third rail, to enable American trains to pass over their line. The section of the Main Trunk east of Montreal had been commenced with a "broad gauge," and that circumstance may have had some influence in determining the decision of the Committee. Erastus Corning, a name influential among railroad men, gave his opinion in favor of the four feet eight and a half, to enable our roads to connect with railroads in the States, which had adopted that gauge; the New York, Northern and Central, and the New England lines. And he held that, not one advantage to a wide gauge can be stated without a sacrifice incident to such increase." At the same time he stated with great candour, "that the relative advantages and disadvantages of various gauges rest solely upon the stability of the road bed to sustain the weight of engines and cars, and their action when in motion on the track." Another competent witness stated the difference between a passenger car, broad or narrow, at $200 to $250. H. C. Seymour, State Engineer of New York, admitted the inconvenience of a gauge that necessitated transhipment; but he contended that all the objections to a five and a half feet gauge had been refuted by the result of actual experience. "Besides the decreased wear and tear consequent upon the easier motion of the cars and engines on a wide gauge," he said, "the comfort of passengers produced by the wider seats permissible in cars running on a wide gauge, is an important consideration." A five feet and a half track would enable the cars to be a foot wider than on one four feet eight and a half. John A. Roebling, civil engineer, adduced the fact that the relative number of accidents on the two gauges showed the narrow gauge to be the safer of the two. And he explained how this difference might be accounted for :

"Trains generally run off on curves. and as these can never be altogether avoided, but only reduced at an outlay of capital, their effect, as influenced by the gauge, has to be principally considered. The wheels of locomotives as well as cars, being fixed stationary upon the axles, and occupying parallel planes, have a tendency to maintain a straight course under all circumstances. When forced, therefore, to move around a curve, the outer wheels, rolling over larger space than the inner ones, are forced to slide to make up the difference. But this sliding cannot be effected without meeting a great resistance, which is equal to the adhesion between rail and wheel, resulting from superincumbent pressure. This resistance is aided by the natural tendency of all moving bodies to preserve a straight course, which is the tangent of the curve. When, therfore, these forces, tending towards the preservation of the straight line, are greater than the resistance of the flanches acting against the sides of the outer rail, and perhaps aided by some small obstruction or inequality on the track, the consequence will be a run off. Now the strife between the inner and outer wheel increases with the width of track ; therefore, the narrower the track the greater the safety. The conical shape of the tire has been found to avail but little, and is nearly abandoned. On the other hand the steadiness of cars moving around curves is more insured by a wide guuge than by a narrower one."

Still Mr. Roebling advocated a width of gauge from five feet three to five feet eight and a half, on account of its allowing the construction of cars of greater width. This was on the supposition that there were no controlling reasons for the adoption of any other gauge. But when he had to answer what gauge would, in existing circumstances, be the best for the Canada trunk line, he said :

"If these lines (Great Western, Toronto and Hamilton, and others east) are to form a great system in themselves, self-supporting and independent of others, I should adopt a gauge of five feet three inches. If its connection with the Portland Road, which has a track of five feet six inches, is of any great importance, I should adopt the latter. The position of the Great Western line, however, appears to me a different one. This can never be exclusively a Canadian line, it will be more an American one, as it will form one of the most important links in the great route from Boston to Chicago, the great parallel rival of the New York and Erie. To attempt to make it a Provincial Line exclusively, would be destroying its future prospects, and reducing its support to the local travel and traffic, which, for a number of years will be insufficient to maintain a good line. Canada West is intermediate ground between Michigan and the Great West on one side, and New York and the Eastern States on the other. A change of gauge at the frontiers would, therefore, be bad policy. A large portion of produce and live stock raised in Michigan will seek this route, and no change of cars should take place. Transhipment of freight and live stock is expensive, and causes delay, and should by all means be avoided on the run from Michigan to Albany. If a wider gauge is considered preferable for the Lower Canada Line, the track of the Great Western should correspond with that of the Niagara, Lockport & Rochester line."

Mr. Thos. Rodgers, of Paterson, New Jersey, who had had great experience in building engines, and was besides a heavy stockholder in narrow gauge railways, mentions several practical objections to the four feet eight and a half track. The demand for increased speed that had sprung up, made it very difficult to put in a boiler sufficiently large to generate steam enough; and it was very difficult to arrange the different parts of the machinery properly without raising the boiler higher than desirable. The necessity of raising the engine high, causes it to roll much more, in going round a curve, than on a wider gauge, and much more weight is thrown on the outer rail, causing increased friction, and wear and tear, with a loss of power at the very time when the greatest power is required. For these reasons he thought it would be possible to take a much heavier train over a five feet and a half gauge than one of four feet eight and a half. Improvements have now been made in the engines, by which they are made to sit close to the ground, so that the above objection no longer exists. Often the flues had to be made so long that their expansion and contraction rendered it impossible to keep them tight. In case of a five and a half feet track none of these difficulties were presented. Mr. Killaly, then attached to the Public Works Department as engineer, recommended a five feet six gauge, partly on the ground that several miles of what must comprise part of the Main Trunk, were already built of that width. He held that it would give greater safety than the narrow gauge, but he certainly did not answer the objections of Mr. Roebling. There was an advantage in obtaining larger driving wheels, in decreasing the velocity and friction of the piston and in the more free and easy working of the engine. He laid it down as an axiom that the more tonnage of net freight the engine can draw, the less in proportion will be the cost of running. Still he did not consider there was any such difference of superiority of one of these gauges over the other, that he would be justified in deciding the question on their abstract merits. And consequently his decision was based upon the necessity of connecting the other sections with that on which the five feet six tracks had been laid. He totally rejected all arguments drawn from the desirability of connecting with the New York lines; being fully convinced of what we now know was an error, that a change of cars would always take place at the frontier.

With all this evidence before them, and all these circumstances to be considered, the Railway Committee, on the 31st July 1851, decided in favor of the five feet six gauge. The resolution which embodied this decision was moved by Mr. (now Sir) J. A. Macdonald, and it authorized the Government to recommend to the Directors of the Great Western Railway to adopt this gauge. In the vote on the question, the Committee stood nine against two. In this way, what has since been known as the Provincial gauge came to be adopted.

INTERPROVINCIAL NEGOTIATIONS.

In compliance with a suggestion of Earl Grey, in his despatch of March 14, 1852 a deputation from the Governments of Nova Scotia and New Brunswick, visited Toronto, to confer with the Canadian Government, "for the purpose of coming to some agreement, on the subject, which, after being approved by the Legislatures of the several Provinces, might be be submitted for the sanction of Parliament." Mr. Howe represented Nova Scotia and Mr. Chandler New Brunswick. They reached Toronto on the 15th June. New Brunswick, though thus represented, was still hesitating ; and all that could be done by the Conference was to agree upon a basis of action to be submitted to the Government of that Province. That basis at once dissipated the illusions which the Canadian legislation of that year created. It was agreed, subject to the approval of New Brunswick, that the line from Halifax to Quebec should be made " on joint account and at the mutual risk of the three Provinces, ten miles of land along the line [on both sides it is to be presumed being voted in a joint commission, and the proceeds appropriated towards the payment of the principal and interest of the sum required." New Brunswick was to construct the Portland line—the North American and European—at her own risk, with funds which it was erroneously assumed would be advanced by the British Government, while Canada, at her own risk, was to build the line between Quebec and Montreal, and any saving that could be effected out of the share of the Halifax and Quebec Railway guaranteed loan, was to be appropriated to the extension of the line above Montreal. When the debt contracted, on the joint account of the three Provinces, should be repaid, each Province was to own the portion of the line within its own territory. Canada was to withdraw the general guarantee offered for the construction of railways in any direction, and her resources were to be concentrated upon the main line, with a view to the early completion of a great intercolonial and interior highway from Halifax to Hamilton ; thence to Windsor, opposite Detroit, the Great Western, then in course of construction, was to complete the line to the Western frontier of Canada.

The New Brunswick Government agreed to accept these terms, as soon as assured that it had been confirmed by that of Nova Scotia. Mr. Howe, in his arguments to obtain this confirmation from the people of Nova Scotia, who were about to elect a new Legislature, even then argued that this line would in our time, be extended to the Pacific. All the calculations were based on the assumption that the railway would cost £7,000 currency or $28,000 a mile ; but Mr. Howe thought that much of the work could be done for $20,000 a mile. He found that the capital with which American railroads had been constructed had cost from seven to twelve per cent. ; and he brought his mind to the conclusion " that a railroad built with money at 3½ per cent., will pay almost immediately, even if made through a wilderness, provided the land be good, water power and wood abundant ; and provided there are settlements at either side, to furnish pioneers and local traffic with them when they are scattered along the line." This is a more hopeful view than most persons now venture to take of the Intercolonial. Mr. Howe estimated the quantity of land to be appropriated in aid of the railway, chiefly by Canada and New Brunswick, at three million of acres; and argued that if it were sold at a dollar an acre it "would form a fund out of which to pay the whole interest on the capital expended for the first three or four years."

It afterwards appeared that Mr. Howe was mistaken in supposing that the promised Imperial guarantee was to extend to the North American and European Railway, from the straits of Cumberland in the east, to the boundary line of the United States in the west. At least Sir John Pakington, now become Colonial Secretary in place of Earl Grey, so construed the letter of Mr. Hawes, to which reference has been made. The difficulty arose on a single sentence. "Her Majesty's Government will by no means object to its [the Intercolonial Railroad] forming part of the plan which may be determined upon that it should include a provision for establishing a communication between the projected railways and the railways of the United States." But it was one thing for the Intercolonial to have a connection with American railways, and another thing for the British Government to guarantee the capital with which to build that connecting link. But Sir John Pakington stood on less secure ground when he argued (despatch to the Earl of Elgin, May 20th, 1852) that no pledge had been given of assistance to any line except that originally proposed"—Major Robinson's line. Mr. Hawes had been very explicit on this point. "Her Majesty's Government do not require," he said, "that the line shall necessarily be that recommended by Major Robinson and Captain Henderson." Sir John Pakington insisted on Major Robinson's line, and refused the guarantee to one running in the valley of the St. John. The transference of the line, by the Provinces, from the north shore to the valley of the St. John, had arisen from a desire to accommodate the difficulty which had sprung up about the guarantee to the North American and European Railway, and when this compromise failed to meet the views of the Imperial Government, Canada and Nova Scotia each commenced the construction of leading lines on their own account and on the strength of their own unaided credit. And New Brunswick set about the construction of the North American and European Railway. The Imperial pledge of a guarantee to the Intercolonial, not absolutely withdrawn but only refused to a particular line, was one day to be fulfilled, with that scrupulous fidelity that attaches to all the obligations of Great Britain.

When Sir John Pakington announced the refusal of the Imperial Government to extend the guarantee to the line along the valley of the St. John, Mr. Hincks and Mr. Chandler had gone to England to arrange the details of the agreement to be completed. Mr. Howe, the third delegate, had not yet arrived when they had their first interview with Earl Derby, the new premier. They were promised another interview on the arrival of Mr. Howe. But Mr. Hincks, on the very next day, May 1, 1851, addressed a letter to Sir John Pakington, in which the failure of the negotiations was anticipated, and a new line of policy to be followed, in that event, pointed out. If he did not obtain a final answer by the 15th—in fourteen days—he should on the part of

Canada withdraw from the negotiation; as he had reason to believe that he could effect arrangements with eminent capitalists to construct all the railroads necessary for Canada, on the unaided credit of the Province." Such a statement seems ill calculated to secure the Imperial guarantee.

CONTRACT WITH PETO AND CO.

By the 9th of the month Mr. Hincks had had several personal interviews with Mr. Jackson on the subject of a contract. The substance of these conversations was that Messrs. Peto, Brassey, Betts and Jackson were to undertake the construction of a railroad from Montreal to Hamilton, at a rate which would, by their own estimate, produce them the same profit they had made in England and on the Continent of Europe. On the twentieth, the negotiations began to be reduced to writing, and the next day a basis of agreement was arrived at. The contractors were to send out engineers to survey the line, and if any difficulty occurred, the Government was to pay the cost. To the extent of five-tenths of the capital, the direct bonds of the Government were to be issued instead of the company's bonds guaranteed by the Government. They were to be issued through Baring Bros., and Glyn, Mills & Co., "to whom," Mr. Hincks said, "the Canadian Government is bound not to allow its bonds to be issued through other parties."

NEW RAILWAY LEGISLATION.

This agreement involved a new policy of railway legislation. But before coming to what that legislation was, we must first recapitulate what had been previously done on some sections of what was now to be called *The Grand Trunk Railway of Canada.*

In 1848, the *Toronto and Goderich Railway Company* was chartered, (10 & 11 Vic. cap. 123) with a capital of £750,000, in shares of £25 each, with power to raise an additional sum of £250,000 if required. This road, in its passage from Toronto was to strike Guelph and the waste lands of the Crown lying north of the Huron Track, to Goderich on Lake Huron. The survey map and book of reference were to be deposited within three years and the road to be completed within ten years. Construction was not to commence until £150,000 of the stock had been subscribed, and ten per cent. paid on it. The Directors were empowered to unite with any joint stock company then formed or to be hereafter formed in the United Kingdom, and with the Toronto and Lake Huron Railroad Company.

In 1851, *the Kingston and Montreal Railroad Company* was incorporated, with a capital of £600,000 currency ($2,400,000), in shares of $100 each; and if that proved insufficient, power was given to raise £400,000 more. The same power of making arrangements as in the old act was given. The gauge was fixed at five feet six inches. The whole of the stock was subscribed by ten persons, in August, 1852.

Date	Signatures	Residence	No. of shares	Amount £ s. d.
1852 August [illegible]	J. Torrance,	Montreal	[illegible] shares	[illegible] 0 0
"	William [illegible],	Do	[illegible]	[illegible] 0 0
"	John [illegible],	Do	[illegible]	[illegible] 0 0
"	H. N. Jones per Attorney John Young,	Do	[illegible]	[illegible] 0 0
"	[illegible],	Do	[illegible]	[illegible] 0 0
"	William [illegible],	Do	[illegible]	[illegible] 0 0
"	Thomas Galt per Attorney A. T. Galt,	Toronto	[illegible]	[illegible] 0 0
"	A. T. Galt,	Sherbrooke	7,500 shares,	[illegible] 0 0
"	L. H. Holton,	Montreal	7,500 shares,	[illegible] 0 0
"	D. L. Macpherson,	Do	7,500 shares,	[illegible] 0 0
				[illegible] 0 0

By articles of preliminary agreement, each of the three largest subscribers bound himself not to transfer any of the stock, without the express authority of the other two in writing. When the question of chartering the Grand Trunk, which would cover the same ground as this company's charter, came up, the Railroad Committee refused the petition by the Kingston and Montreal Railroad Company, against the proposed new charter, on the ground of this very agreement, which was construed into proof that the subscription was not *bona fide*. Another reason was that Mr. Jackson, on behalf of Peto, Brassey, Betts & Co., had offered to construct the whole line, and be responsible for the floating of all the stock of the Grand Trunk, on obtaining the Government guarantee of £3,000—$12,000—a mile. Mr. Holton, Mr. Galt and some others, acting as promoters of the Kingston and Montreal Co., had employed Mr. Gzowski to make a survey; and at the time this parliamentary contest [illegible], their expenditure was set down at £800, besides the cost of the survey made by T. C. Keefer, which was nearly $8,000. The Committee asked them to resign their stock, and accept repayment of their outlay. After much skilful fencing between Mr. Holton and Galt, on the one side, and the Railway Committee, and Mr. Hincks, on the other, in which scarcely any thing was said about the real thing in dispute—the railroad contract—Mr. Holton, finding the enemy too strong for him, offered to capitulate on the following terms: That assurance should be given by the Government that they could get the road built, as well as a bridge over the St. Lawrence, without increasing the guarantee over £3,000 a mile, and that the use of such bridge should be secured to all railroads. The preliminary expenses of the Company were to be repaid, and their liabilities assumed by the Government. Mr. Hincks made some objection to making the immediate construction of the bridge a condition, and an arrangement was concluded from which this item was left out. The rest of the terms were as stated. The contest was afterwards renewed, but the Government had the control of the guarantee in its hands, and the Railroad Committee seconded Mr. Hincks's proposal to let the contract to English contractors.

The reasons for letting the Grand Trunk contract to Jackson, Peto, Brassey and Betts, was their alleged ability to float the stock. The Quebec and Richmond Railway Co., had let a contract to this firm, at £6,500 a mile. Before doing so, their agent, Mr. Wm. Chapman, who was a director of the Bank of British North America, in London, had for months tried in vain to float any of the stock; but after this contract was given he succeeded in a short time in selling £205,000 of stock and £100,000 of bonds at par. Mr. Young was of opinion that if the contemplated arrangement with Peto, Brassey & Co., were broken off, the Grand Trunk could not be built for many years to come. Mr. Galt and Mr. Holton continued to be of a different opinion; and they alleged that with a provincial guarantee of £2,500 a mile, they could construct the Kingston and Montreal section in less time than it would otherwise be constructed, and with a much less amount of stock. "We ask" they said "no power to issue excessive amounts of stock—deluding strangers into the belief that works are costly which are really cheap."

The *Act to Incorporate the Grand Trunk Railway of Canada* (16 Vic., Cap. 37), passed in 1852, incorporated a company with a capital of £3,000,000 stg., in £25 shares, to construct a railway, on a designated route, from Toronto to Montreal. The Government guarantee, to be given in the form of Provincial debentures, was confined to £3,000—$12,000—a mile, and was to be handed over in amounts of £40,000, whenever £100,000 stg. should be ascertained to have been expended "with due regard to economy" on the road.

Another Act, (Vic. 16, Cap. 38) was passed the same session, *To provide for the Incorporation of a Company to construct a Railway from opposite Quebec to Trois Pistoles, and for the extension of such railway to the eastern frontier of this Province.* The capital was fixed at one million sterling, with power to increase it to four millions, and the right to extend the road to the eastern limit of the Province. The same amount of Provincial guarantee as in the case of the Grand Trunk was to be given to that section which lay between Point Levi and Trois Pistoles; but for an extension a grant of a million acres of land was to be given in lieu of a money aid. In other respects the terms of this Act were the same as those of the preceding.

What is popularly known as the *Amalgamation Act* (16 Vic., Cap. 39) completed the series of railway legislation this session. It empowered any railway company whose road formed part of the Main Trunk line to unite with any other such company. Its provisions were applied to the St. Lawrence & Atlantic Railway Co., and the railway which that company was empowered to construct. It repealed the Acts incorporating the Montreal & Kingston Railway Co., and the Kingston & Toronto Railway Co., and obliged the Grand Trunk Railway Co. to pay the promoters of these railways the preliminary expenses they had incurred.

In 1853, the Grand Trunk Railway Company was authorized to increase its capital or to borrow to the extent of £1,500,000 sterling, for the purpose of constructing a general railway bridge across the St. Lawrence at or in the vicinity of Montreal. It might undertake the work alone, or in conjunction with any other company or companies. The plan was to be approved by the Governor in Council. This bridge is one of the noblest monuments of engineering skill the world has ever seen. Its total length is 9,184 lineal feet (nearly a mile and three-quarters). It has 25 spans of tubular iron, 24 of which are 242 feet each, and the other 330 feet. These are 60 feet above the water level, and rest on a series of piers which contain altogether 3,000,000 feet of masonry. The centre pier is 24 feet wide; the others 16. They have to resist at all times a current of seven miles an hour, and, in the spring, immense masses of ice, which have had no injurious effect on them. The work was commenced on the 20th July, 1854, and a passenger train passed over the bridge December 17, 1859. This stupendous work was originally designed by Mr. Stevenson and Mr Ross. The general design was matured and worked out by the latter.

By another Act, passed the same session (Vic. 16, cap. 76), the Amalgamation Act was extended to companies whose railways intersect the main trunk or touch places which that line touches. In pursuance of this Act, the Toronto and Sarnia, the Toronto and Kingston, the Quebec and Trois Pistoles, and the Belleville and Peterboro'—the latter a projected branch which was never built—were united. The negotiations were conducted in London in the first five months of 1853; Mr. Galt representing the Atlantic and St. Lawrence, the St. Lawrence and Atlantic, and—in connection with Mr. Alexander Gillespie, of London—the Toronto and Guelph railway companies, Mr. Ross, the Grand Trunk proper, as its President, and the eastern section of that road, in connection with Mr. Forsyth and Mr. Rhodes.

The amalgamated company assumed all the liabilities of the several companies, which, previous to the amalgamation, had a separate existence. This included a contract with Messrs. Gzowski & Co., entered into on the 24th March, 1853, for the construction of the Toronto and Sarnia section, for the sum of £1,376,000 sterling, the distance being estimated at 172 miles; Messrs. Peto, Brassey, Betts, and Jackson's contract, entered into one day before Gzowski & Co.'s was signed, for the construction of the line between Montreal and Toronto, estimated at a distance of 345 miles—eleven miles over the real distance—for the sum of £3,000,000 sterling; the contract with the same parties, dating October 20, 1852, for the construction of the Point Levi and Richmond line, some 95 miles, for the sum of £850,000; a contract with the same parties for the construction of the Quebec and Trois Pistoles road, estimated at 153 miles, for the sum of £1,224,000 sterling; a contract with the same parties, never executed, for the construction of the Belleville and Peterboro' line for the sum of £400,000; and a contract with the same parties, executed March 3, 1854, for the construction of the Victoria railway bridge at Montreal, for the sum of £1,400,000 sterling. The Atlantic and St. Lawrence Company, whose road runs from Portland, Me., to Island Pond, Vt., a distance of 148 miles, leased its property to the Grand Trunk for a period of 999 years, at a yearly rent representing six per cent. on the share and stock capital, $1,700,000, besides the interest on the bond and debenture debt; in all, $2,000,000 a year, payable half-yearly on the 1st January and the 1st July.

CAPITAL STOCK.

The prospectus of the Grand Trunk Railway was issued while the arrangements for a fusion of the companies were in progress, under the guarantee of powerful names of the monetary world of London and seven members of the Executive Government of Canada. Among the London Directors were Baring, representing one house, and Glyn another, and both of them were members of the House of Commons. The Government directors in Canada were the Hon. John Ross, Solicitor General for Upper Canada, Hon. F. Hincks, Inspector General, Hon. E. P. Tache, Receiver General, Hon. Jas. Morris, Postmaster General, Hon. Malcolm Cameron, President of the Executive Council. Glyn, Mills & Co., and Baring Bros., were the bankers, and Alexander Ross was engineer in chief. The prospectus contained the following statement:

"The capital is.. £9,500,000
made up as follows:

"Amount already raised in shares, and spent on works of the St. Lawrence and Atlantic and Quebec and Richmond Railways..........	£ 683,400	
"Amount already raised on bonds..............	733,000	
	£ 1,416,400	
"Received in shares and debentures for the shareholders in the St. Lawrence and Atlantic, and Quebec and Richmond Railways, on the amalgamation, and for the bondholders of the Ontario, Simcoe and Huron Railway Company....................	£ 837,600	
		2,254,000
"Leaving..		£7,246,000

"This amount will be created and apportioned as follows:

Stock in 144,920 shares of £25 each..............................	£3,623,000
Debentures of £100 each, payable in 25 years, bearing interest at 6 per cent. per annum, payable half-yearly, in London, and convertible into shares on or before the first day of January, 1863, at the option of the holder	1,811,500
"And debentures convertible into bonds of the Provincial Government, of £100 each, payable in 20 years, bearing interest at 6 per cent. per annum, payable half yearly in London....................................	1,811,500
	£7,246,000

"Of these 144,920 shares it is proposed now to issue one-half, viz.: £1,811,500 in shares, and the same amount in debentures, the other half having been agreed to be taken by the contractors, who, however, engage to give to the holders of such shares, on the 1st July, 1854, (twelve months after the anticipated opening of the St. Lawrence and Atlantic section of the railway) the option of taking, in equal proportions, two-thirds of such remaining moiety; that is to say, every holder of thirty such shares will, on the 1st July, 1854, be entitled to claim twenty shares more at par, together with an equal amount of debentures, also at par. Such additional shares and debentures to bear interest at 6 per cent. from the said 1st July, 1854.

£400 of debentures (one-half of each description) will be issued at par, with each £600 of shares.

"By the law granting the Provincial aid, it provided that the bonds of the Province shall be issued as the works advance. These bonds, will, therefore, be held in trust to be delivered pro rata to the holders of the convertible debentures."

The estimated profit was nearly 11½ per cent. The gross estimated earnings have been fully realized; but the great error of calculation, which makes all the difference between profit and loss, was in putting down the working expenses as low as forty per cent., the actual amount having been from seventy to eighty per cent.

The description and objects of the Grand Trunk Railway are fully set forth in the appendix, to which especial reference is craved.

The more prominent points therein are:—

1. The completeness of the system of railway, engrossing, as it does, the traffic of Canada and the State of Maine, and precluding injurious competition.

2. The large amount of government guarantee and of Canadian capital invested, being two millions eight hundred thousand pounds sterling.

3. The fact that 290 miles of the railway are now open for traffic, to be increased to 590 miles by the close of the present year.

4. The execution of the whole remaining works being in the hands of most experienced contractors, the eminent English firm of Messrs. Peto, Brassey, Betts and Jackson having undertaken seven-eighths thereof, including the St. Lawrence Bridge.

5. The cost of the railway being actually defined by the contracts already made, whereby any apprehension of the capital being found insufficient is removed.

In the appendix will also be found the data for the following summary of probable revenue:—

On 1112 miles, at an average of above £25 per mile per week	£1,479,660	
Deduct working expenses, 40 per cent	591,864	
		£887,796
Interest on Debenture debt, £4,685,200	278,100	
Rental of Atlantic and St. Lawrence Railway	60,000	
		£338,100
Thus showing a profit on the share capital, £4,864,000, of nearly 11½ per cent		£549,696

There ought surely to have existed data on which more accurately to estimate the cost of the working expenses. The following statement of the distribution of the capital over the different sections was appended to the prospectus :

ORIGINAL *distribution of Capital.—Estimated cost of several works comprised in the Grand Trunk Railway Company :*

St. Lawrence and Atlantic, 142 miles—£8,500 per mile...		£1,258,000
Quebec and Richmond, 100 miles—£6,500 per mile—......	£650,000	
Extension...........................	50,000	
		700,000
Montreal to Toronto......................		3,000,000
Trois Pistoles, 153 miles, at £8,000 per mile............................		1,244,000
Grand Trunk Junction, 50 miles, at £8,000 per mile.................		400,000
Toronto and Sarnia, 172 miles, " "		1,376,000
Victoria Bridge		1,400,000
Contingencies		142,000
Sterling...		£9,500,000

The issue of the first half of the Stock, £1,811,500, in £25 shares, was attended with surprising success. The applications were immensely in excess of the amount to be issued—some put the whole amount applied for as high as twenty millions sterling—and brokers speculating in the stock, in advance of its issue, agreed to deliver shares at £1 premium. There was naturally great disappointment among the applicants ; a feeling that was not to be without its compensation in the future. The stock issued at par went up as high as two per cent. premium ; but when it once fell below par it never recovered, but steadily declined till quotations became merely nominal.

It would seem that a great mistake was made in not issuing the whole of the stock at once ; for that was the only time when it could have all been floated at par. But this could not have been foreseen, at the time.

The Provincial guarantee extended to the various sections of the road, in the following proportions, amounted to £1,811,500 stg., to be represented by six per cent. debentures, payable in twenty five years, and to be issued on the conditions previously stated :

Toronto to Montreal.	345 miles.
Quebec to Trois Pistoles..................	153 "
	498 miles.

At £3,000 per mile...	£1,494,000
St. Lawrence and Atlantic	67,500
Quebec and Richmond......................	250,000
	£1,811,500

Besides this, £400,000 had already been issued to the St. Lawrence and Atlantic Railroad prior to the amalgamation. The whole amount thus authorized by the Legislature to be issued was £3,111,500 stg. It was seriously argued that by agreeing to issue this amount of debentures in aid of the Grand Trunk, the Province was "only incurring a nominal responsibility;" this was admitted, Mr. Hincks said, in 1868, even by the opponents of the bill; the idea being that the first mortgage held by the Province constituted ample security for the advance.

The amalgamation was confirmed by the Legislature in 1854, and transferred to the amalgamated company the title of the Grand Trunk Railway of Canada. The company was authorized to increase its capital, but the delusion about securing for the Province its first lien was kept up. The lien, being considered a very valuable thing, was extended to the whole Grand Trunk Railway and its works, and the engineer's certificates for the forty per cent. of guarantee were to extend to all the company's works, though there was to be no increase in the total amount to be issued. No more Provincial aid was to be granted to the Point Levi & Richmond or the Montreal & Portland sections beyond the £717,300 already issued; none to any branch railway that might be thereafter built, or to any line that might be amalgamated with it, except the direct line between Trois Pistoles and Sarnia. The amount of Provincial bonds, that might be issued in aid of the Victoria Bridge was limited to £100,000 stg.

In the early part of 1855, efforts were made to obtain for the Company additional assistance from the Canadian Government. The English contractors wrote to Mr. Thos. Baring and Mr. George Car Glyn, both of whom occupied the double position of directors of the Company and financial agents of the Government, stating at what rate they would push on the work of construction, if £900,000 of additional Provincial aid were obtained. They would open the road from Montreal to Brockville, and from Quebec to St. Thomas, in the ensuing autumn; they would open the additional section from Brockville to Toronto, giving a railway connection between Montreal and Toronto by the autumn of 1856. This promise was left a year behind in the performance. But the line from Quebec to Richmond was (February 2, 1855) already open, though the contract gave them over ten months more, and a year over the time when the road had been opened (December, 1855.)

This appeal was responded to. In the latter part of the session of 1854–55, an Act was passed, (18 Vic. Cap. 174), May 19, 1855, to grant additional aid to the Grand Trunk Railway. It authorized the issue of Provincial debentures to the amount of £900,000 stg., redeemable in twenty years, for this purpose. The conditions on which they were to be issued to the Company were that the whole amount of aid received and to be received, for work or materials on the general should not exceed fifty per cent. of the whole amount expended on the work, prior to the 1st May, 1855; and the sum to be advanced out of this additional grant was never to exceed seventy-five per cent. on the amount expended by the company, after that date, on the portion of the line between St. Thomas and Stratford, exclusive of the work on Victoria Bridge. This loan, like the first, was made a first lien on the Company's works; and as the Victoria Bridge, on account of which no Provincial aid was advanced, was included in the mortgage, it was argued that the Province was increasing its security so much that the additional grant was, for it, a good operation, and one which on financial grounds, it would have been madness not to have gone into. The loan was repayable in twenty years, and the interest, six per cent., half yearly. In 1853, 1854 and 1855, while the capital account was in its best condition, the Company did pay interest on Government bonds to the amount of about £200,000 stg. Evidently motives of policy made it advisable for the Company to hold out a prospect that such interest would continue to be paid, as long as additional grants were likely to be required.

But the time was fast approaching when the idea that the lien which the Government held on the works would ever be the means of bringing back the capital advanced, must cease to be entertained by even the most sanguine. In 1856 (July 1,) an Act (19 and 20 Vic., c. 111,) was passed which exploded the idea, advanced a few years before, that the Province only incurred a nominal responsibility in giving the Provincial guarantee to this great national undertaking. The first lien, which had been relied upon as a means of securing the repayment of the capital advanced to the Company, was given up. By the terms of this Act, which had been provisionally agreed to in advance between the Government and the Company, the latter was authorized to issue preferential bonds to the amount of £2,000,000 stg.; these securities to have priority over the Province lien. The issue was not to take place till the railway from St. Thomas to Stratford had been finished and in operation. The proceeds of the bonds were to be deposited with the Provincial agents, in London, and released to the Company on certificates of the Receiver-General, during the progress of the following works:—

The railway from St. Mary's to London and Sarnia....................	£450,000
The railway from St. Thomas, Lower Canada, to Rivière du Loup.	525,000
Victoria Bridge..	800,000
Three Rivers and Arthabaska	125,000
To enable the said Company to assist the Port Hope, and Cobourg and Prescott Railways as subsidiary lines.........................	100,000
	£2,000,000

For the ensuing five years, the time estimated to be necessary for the completion of the construction, the Province was to pay interest on the bonds it had issued in aid of the work; but still the idea of repayment, though in a new form—in the share capital of the Company—was kept up in this Act; and the lien of the Province, subject to these conditions, was to rank, as to dividend or interest, with that of the Company's bondholders.

In this year, 1856, the Company asked the Government to guarantee five per cent. interest on the share capital, but the proposition was not entertained.

On the formation of the Grand Trunk Company, and the grant to it of the Provincial guarantee, it was deemed expedient to give the Government a representation in the direction, with the idea that the interests of the Province would thereby be better guarded. This arrangement was made the occasion of attacks on both the Government and the Company, in which the latter was declared to be too much under political influence. A cry for the abolition of the Government directorate was set up. This would of itself probably not have led to any result, but when the Government lien had been virtually given up, there was no longer any object in retaining the Government Directors. Accordingly, in 1857, there was proposed an Act (20 Vic., c. 11) *To dispense with Government Directors in the Grand Trunk Railway of Canada, and to facilitate the completion of the Company's works from Rivière-du-Loup to Sarnia.* The Government Directors were to go out of office at the next general meeting of the shareholders, and all the powers of the Company were henceforth to be wielded by the elected Directors. The existence of Government Directors in the early years of the Company's existence was afterwards, in 1861, sought to be made, by a committee of the bond and stockholders, the basis of a financial responsibility which the Province had never contemplated and never could be induced to assume. By the Act of 1857, a year's extension of time for completing the works was given, and as a condition of their being completed even within that time, and so long as they are worked and regularly maintained, "the Province foregoes all interest on its claims against the Company, until the earnings and profits of the Company, including those of the Atlantic & St. Lawrence Railroad Company, shall be sufficient to defray the following charges :—1. All expenses of managing, working and maintaining the works and plant of the Company. 2. The rent of the Atlantic & St. Lawrence Railway, and all interest on the bonds of the Company exclusive of those held by the Province. 3. A dividend of six per cent. on the paid up share capital of the Company, in each year in which the surplus earnings shall admit of the same ; and then in each year in which there shall be a surplus over the above-named charges, such surplus shall be applied to the payment of the interest on the Province Loan accruing in such year. The bonds and share capital herein mentioned shall be held to include and consist of all loans and paid up capital which the Company have raised or may hereafter raise *bona fide* under the authority of any Act of the Provincial Legislature, passed or to be passed, for any purpose authorized by any such Act." This was equivalent to a complete surrender of the Provincial lien, and it would have been better to wipe it out altogether than to foster the delusion that anything could in any remote contingency be realized from it.

Next year, 1858, came *An Act* (22 Vic. Cap. 52) *to amend the Acts relating to the Grand Trunk Railway of Canada.* It gave authority to the Company to issue additional bonds, preferential or otherwise, with the now absolutely ridiculous proviso that the new issue should in no way affect the Province lien on the road. And there was a clause providing, among other things, in the nature of priorities, the order in which the interest on the *Provincial* debentures should be paid by the Company. Authority was also given to alter and enlarge the conditions of the lease with the Atlantic and St. Lawrence Railway consistent with the preservation of the relative positions of the Province and the Company.

In 1861, a committee of shareholders drew up a statement in which they asserted "that it was in *bonâ fide* reliance upon the representations put forward as from the Canadian Government in this [the Company's] prospectus, that, in 1853, the petitioners and other persons became subscribers to the Grand Trunk Railway, and in the full persuasion that a Colonial Government which had sought assistance in England in a form so public and conspicuous, would at all times be ready to extend to the obligations thus incurred, at a distance of three thousand miles, not a construction resting on narrow rules of law, but an interpretation large, liberal and statesmanlike," and that they relied on the Canadian Parliament to fulfil this expectation. This was equivalent to asserting that the undertaking was set on foot as a Government work; an assumption which the Canadian Legislature was not likely to endorse. If the Government had undertaken the construction of the road as a public work, the committee argued, it must have incurred an expenditure of £11,000,000 stg., or £660,000 a year, whereas, by the mode adopted, the Province had obtained all the advantages of the Grand Trunk system at a charge of not more than £3,111,500, or £187,000 a year, from which amount there were several deductions to be made. They argued that the Arthabaska branch, which they state at 30 miles, and nearly the whole of the 358 miles forming the Eastern Division, though valuable to the country, must be worked either at a positive loss, or upon terms which will not yield any profit upon the capital expended in their construction; that this is true, in the most unqualified sense, of the 148 miles between Quebec and Riviere du Loup and of the Arthabaska branch, and to some extent of the 96 miles between Richmond and Quebec. They sum up by saying that, as regards the 214 miles east of Richmond, and as regards the branches, the Grand Trunk has become charged with the burden of constructing, maintaining and working lines of railway, not for the benefit of the share and bondholders, but wholly for the present and future benefit of particular portions of Canada; that an amount nearly equal to two-thirds the whole Provincial aid was expended on works valuable to the country, but unprofitable to the Company leaving only £1,111,500 contributed to what they call the commercial portion of the undertaking. It was contended that these facts, all taken together, gave the share and bondholders not a legal, but a strong moral claim on the Province. They estimated the increased market value conferred on the grain and other crops of the Western portion of the Province by the Grand Trunk railway, as not less than 20 to 30 per cent, a statement of which it would require a close examination of a history of prices and other data to test the accuracy. This attempt to make the Canadian Government a joint partner in the expenditure of fifteen millions sterling, was not responded to in the way the committee desired.

In 1862 the Company claimed additional remuneration for the mail service. This service was represented to be worth, for the ensuing twenty-five yearly, a sum that would capitalize at a million and a half sterling. This capitalization was asked for, and with it authority to raise the further sum of £500,000 stg. to complete, repair and equip the line. The passenger receipts of the Company, it was said, the mileage considered, were very light. The time bills were drawn up, not merely to accommodate the passenger traffic, but also to serve mail purposes. The excessive number of miles run to accommodate the postal service caused the trains to be worked at a heavy annual loss, while in Nova Scotia nothing but accommodation trains are being used, and the load of the train being generally made up to the capacity of the engines, the trains proved remunerative. With the capitalized sum sought to be obtained, the Company intended to compound with its creditors in Canada and England. Hints that the road might possibly be closed were thrown out.

In the next session, *An Act for the Reorganization of the Grand Trunk Railway Company* (25 Vic. c. 56) was passed, giving the Company power to issue postal bonds on the securing of the money it gets in payment of the postal service, besides £500,000 equipment mortgage bonds; the latter operating as a first lien on the Company's property. The effect of this was to place the Government lien still further back. The rate of remuneration to be paid for the postal service performed by the Company was long an unsettled question, on which much correspondence with the Government took place. In 1862, it was resolved to settle the dispute by arbitration; but a change of government taking place, the reference was revoked. In 1865, three commissioners, the late Mr. Wm. Hume Blake, Mr. Justice Day, and Mr. G. W. Wicksteed, were appointed a commission to inquire into and report on the subject. They recommended a rate of ten cents a mile for quick passenger trains, and six cents a mile for mixed trains; which, they added, "cannot be considered too high, when it is considered that the Postmaster-General of the United States pays this same road, between the boundary line and Portland, sixteen cents per train per mile, for a single service, and ten cents per train per mile, for a double service."

The proportion which the working expenses bear to the revenue is mainly determined by two unfavorable circumstances. A large part of the Eastern Division of the road is unprofitable; some sections, such as that between Quebec and Riviere du Loup and the Arthabaska branch, being worked at a positive loss. They are a dead weight on the profitable sections, and tend to make the working expenses of the whole line abnormally high in comparison with the revenue. The other cause that contributes more largely to this result is the necessity of receiving competitive rates for through traffic from the west. These rates are determined by the cost of carrying on the cheapest rival routes. Besides, the easternmost section of the line lies in a more severe climate than any other railway in America, a circumstance which, from the accumulations of snow, adds to the working cost and increases the expense of repairs. The construction of the Intercolonial ought to have a favorable effect on the fortunes of the Grand Trunk.

BUFFALO AND LAKE HURON.

An arrangement was entered into between the Grand Trunk and this Company, respecting the division of their traffic receipts, which received the sanction of the Parliament of Canada. The terms of the agreement were thought, by the Directors of the Buffalo and Lake Huron, to operate against the interests of their Company, and accordingly, after protracted negotiations, modifications and concessions were obtained which practically made a new agreement. This agreement provided for a rent-charge, payable by the Grand Trunk to the Buffalo and Lake Huron Company, in perpetuity, by half-yearly instalments, within two months after the 1st January and the 1st July in each year thus:—For the year ending 1st July, 1869, £42,500; for the year ending 1st July, 1870, £45,000; 1st July, 1871, £50,000; 1st July, 1872, £65,000; 1st July, 1873, £60,000; 1st July, 1874, £65,000; 1st July, 1875, £66,000; 1st July, 1876, £67,000; 1st July, 1877, £68,000; 1st July, 1878, £69,000; 1st July, 1879, and every subsequent year, £70,000. £42,500 per annum of the rent charge is to rank next before the first equipment bonds of the Grand Trunk, and the balance will rank next after the second equipment bonds, which the Grand Trunk were authorized to raise. The ordinary shares of the Buffalo Company to be exchanged, one half, or £615,000, for the like amount of Grand Trunk fourth preference, and the other half, £615,000, for the like amount of Grand Trunk ordinary stock. The £42,500 of the rent charge, payable in 1868-69, was liquidated in Grand Trunk second equipment mortgage bonds at par. This road is now a part of the Grand Trunk system.

The International Bridge, now in course of construction at Fort Erie, is expected to have a most favorable influence on the traffic and profits of the line. It will cost the Grand Trunk a rent charge of £20,000 a year for 28 years, at the end of which the sinking fund of £4,000 a year (part of the £20,000 rent) will leave the bridge in the hands of the Grand Trunk free of charge, but with its tolls still coming in from various sources. The present ferry now costs the Company £16,000 a year, and besides saving this the Grand Trunk will have tolls from the Great Western of Canada and the Erie Railway Companies, so that immediately the bridge is up it is calculated to pay the Grand Trunk well, irrespective of the additional traffic it will be the means of throwing on their line.

CAPITAL ACCOUNT.

The capital expenditure on the different divisions, and over the whole property, up to 31st Dec., 1861, with the total capital expenditure to 30th June, 1870, is shown as follows:

Eastern Division (362 miles)—Engineering, £112,574 13s. 11d.; Works and Permanent Way, £2,637,970 15s. 11d.; Stations, Buildings and Offices, £236,872 1s. 2d.; Miscellaneous Stock, £14,441 10s. 5d.; Electric Telegraph, £6,304 11s. 6d.; General Expenses, £186,031 1s. 11d.—£3,194,244 14s. 10d.

Central Division ([illegible] miles)—Engineering, £76,735 1[illegible]s. [illegible]d. ; Works and Permanent Way, £2,919,461 4s. 3d. ; Stations, Buildings and Offices, £[illegible],-[illegible] 4s. 11d. ; Miscellaneous Stock, £[illegible],783 17s [illegible]d. ; Electric Telegraph, £4,[illegible]1 0s. 10d. ; General Expenses, £150,221 [illegible]s. [illegible]d.—£[illegible],[illegible],050 17s. 4d.

Western Division (19[illegible] miles)—Engineering, £46,391 [illegible]s. 10d. ; Works and Permanent Way, £1,[illegible],311 [illegible]s. [illegible]d. ; Stations, Buildings and Offices, £142,-[illegible] 17s. 10d. ; Miscellaneous Stock, £[illegible],0[illegible] 11s. [illegible]d. ; Electric Telegraph, £[illegible],[illegible] 1[illegible]s. [illegible]d. ; General Expenses, £[illegible]1,013 1[illegible]s. [illegible]d. ; Compensation to Contractors, £25,000 0s. 0d.—£1,811,[illegible]1 7s. 3d.

Portland Division, Leased Line, (149 miles)—Engineering, £[illegible],700 7s. [illegible]d. ; Works and Permanent Way, £192,764 1s. 11d. ; Stations, Buildings and Offices, £74,09[illegible] 1[illegible]s. [illegible]d. ; Miscellaneous Stock, £1,491 1[illegible]s. 3d. ; Electric Telegraph, £1,946 7s. 5d. ; General Expenses, £94,378 [illegible]s. 0d. ; Rolling Stock, £[illegible],[illegible] 14s. 7d. ; Lands in Portland Division, £1,873 7s. [illegible]d.—£[illegible],100 1[illegible]s. 4d.

Rolling Stock, £1,019,791 8s. 11d.

Sundries—Expended on Works, &c., Detroit Line, £4,858 1[illegible]s. 0d. Three Rivers and Arthabaska Branch (Advanced), £106,762, 8s. 10d. Aid to Subsidiary Lines, C. W., £[illegible]7,350 0s. 0d. Port Hope Railway Junction, £[illegible]-14s. 1d. St. Lawrence and Champlain Junction, £[illegible]49 1[illegible]s. [illegible]d. Montreal Extension Survey, £[illegible]16 3s. 1d. Intercolonial Railway, £[illegible] 17s. 11d. Expended on Steam Ferry Boats, Wharves and Barges, £[illegible],957 1[illegible]s. 4d. Buildings, &c., at Sarnia, with Survey, £9,681 11s. 4d. Subscriptions to St. Lawrence Warehouse and Dock Company, £[illegible]5,273 1[illegible]s. 6d. Discount on Sale of Stocks and Debentures, &c., £4[illegible],350 1[illegible]s. 6d. Less premium on sale of Debentures, £67,980 16s. 0d. Expenses of London Office, £[illegible],[illegible]-1[illegible]s. 11d. Victoria Bridge (2 miles), £1,366,0[illegible]0 1[illegible]s. [illegible]d.

Lands and Land Damages, £45,092 6s. 9d.

	£	s.	d.
Total Expenditure on 1,036 miles	£11,960,4[illegible]	12	1
Additional Expenditure to 30th June, 1870	6,634,541	16	3
Total Expenditure	£18,[illegible],0[illegible]	8	4
Unexpended Balance	3[illegible],[illegible]1	18	9
Total	£18,[illegible],510	7	1

PER CONTRA—*Share Stock*—Shares Consolidated into Stock, £[illegible],610,14[illegible]-0s. 0d. ; *Shares not yet Consolidated, £61,648 16s. 5d. ; Received on Shares Forfeited, £1,501 16s. 6d.—£2,77[illegible],[illegible]79 1[illegible]s. 2d.

Debentures—Island Pond Debentures, £60,000 0s. 0d. ; British American Land Company's Debentures, £[illegible]0,547 1[illegible]s. 11d. ; Montreal Seminary Debentures, £[illegible]0,547 1[illegible]s. 11d. Total £131,0[illegible]6 17s. 10d. Mortgage to Bank of Upper Canada, £[illegible]1,100 0s. 0d. Atlantic and St. Lawrence Deferred In-

* Shares in the St. Lawrence and Atlantic Line held by City of Montreal.

terest Certificates, (1872), for arrears to 31st December, 1862, £77,180 11s.-10d.

Preference Bonds and Stocks—Equipment Mortgage Bonds, $500,000. Amount received on do., No. 2, £110,480. Postal and Military Service Bonds, £1,200,000. First Preference Bonds, £2,703,324 16s. 0d.; First Preference Stock, £77,064 4s. 0d.—£2,780,389. Second Preference Bonds, £1,610,264-7s. 5d.; Second Preference Stock, £45,889 12s. 10d.—£1,656,154 0s. 3d. Third Preference Stock, £758,509 17s. 9d. Fourth Preference Stock, £5,-571,120 18s. 3d.

Provincial Debentures—Issued on account of Grand Trunk Railway, £3,111,-500. Amount received on unissued Debentures and Debenture Certificates allotted with forfeited Shares—Company's, £3,650; Provincial, £3,650.—£7,300.—Grand total £18,998,510 7s. 1d.

In 1861 the Line was embarrassed with a floating debt of over twelve millions of dollars, and was absolutely without credit. The condition of the Line too, was such that constant and heavy renewals and repairs have been required to be made every year since. It is necessary to bear these two facts in mind in looking at the Company's present position, in order fairly to appreciate the exertions of its present management to bring it into a state of efficiency.

GRAND TRUNK RAILWAY.

YEARS ENDING JUNE 30th.

ROLLING STOCK, &c.	1861	1862	1863	1864.	1865.	1866.	1867.	1868.	1869.	1870.
Locomotives, Pass No.	80	80	80	83	101	101	101	101	101	107
" Freight No.	143	147	147	171	184	197	197	197	[illegible]	219
Cars, Passenger, 1st class No.	68	88	88	101	131	131	131	131	131	131
" Sleeping No.	9	9	9	12	14	16	16	16	16	16
" Passenger, 2nd class No.	46	46	45	49	64	64	64	64	64	64
" Post Office and Baggage No.	60	60	60	66	77	79	79	85	85	[illegible]
" Box No (incl. cattle cars & [illegible])	1,943	1,943	1,943	2,325	2,391	2,641	2,641	2,661	2,661	2,661
" Platform No	1,058	1,058	1,058	1,107	1,256	1,307	1,307	1,307	1,307	1,307
" Other No.	76	76	76	100	124	101	101	101	101	101
Snow Plows No.	33	33	33	34	34	34	34	34	34	36
Pass. Trains, Mileage, average per month	84,832	71,638	76,674	115134	148,175	140,209	154,024	153,779	154,557	166,444
Freight " " "	117410	135,341	161,328	172579	181,593	201,312	197,304	213,676	237,033	249,143
Mixed " " "	47,709	58,510	51,847	39677	50,907	51,108	45,950	46,043	40,707	62,330
Other Mileage " "	78,019	87,323	98,920	102[illegible]	124,638	129,001	140,911	151,226	148,077	165,817
Total Locomotive Mileage, av'ge per month	327970	352,812	388,767	[illegible]	510,113	530,630	541,169	565,334	579,394	642,772
Passengers, Local, No	......	659,457	617,136	......	1,155,043	1,160,096	1,140,895	1,157,806	1,188,[illegible]	1,207,[illegible]
" Through, No.	....	95,369	114,009	......	263,516	290,818	250,905	231,398	257,815	264,644
" Immigrants, No.	......	13,586	16,837	......	15,621	23,540	28,920	24,330	32,007	31,[illegible]
Freight, Local, tons	......	602,224	533,860	......	737,739	814,911	839,080	814,177	[illegible],130	[illegible]
" Through, tons	......	86,918	100,799	......	118,418	130,929	140,874	152,286	134,[illegible]	163,440
Live Stock, tons	......	6,917	21,162	......	62,496	75,255	42,586	56,509	84,290	101,647
Locomotive expenses per cent on gross rec'pts	24.68	26.93	21.68	19.32	19.40	20.60	22.75	23.05	22.34	22.86
Locomotive expenses, rate per train mile, cts	27.57	30.20	26.40	24.37	25.19	27.29	30.22	31.42	30.41	32.02
Car expenses, cost per train mile, cents	9.30	9.22	8.72	7.67	8.38	8.41	8.74	9.09	9.08	9.02
Car expenses, per centage on gross receipts	7.60	7.91	7.18	6.21	6.46	6.90	6.07	6.88	7.02	7.09
Renewals, per centage on gross receipts	......	......	......	9.62	10.12	8.29	11.20	11.30	14.89	9.60
Total ord'y work'g exp's. " "	85.91	72.98	61.08	59.57	59.55	59.36	64.61	64.71	66.51	66.09

GRAND TRUNK RAILWAY.

FISCAL YEAR ENDING 30th JUNE.

	1866 £. Stg.	1867 £. Stg.	1868 £. Stg.	1869 £. Stg.	1870 £. Stg.
REVENUE.					
Passenger Traffic, Local					
" Through	396,750 18 6	307,723 14 7	307,273 3 0	305,883 9 11	310,816 15 6
" Immigrants	155,829 10 0	160,120 8 7	147,534 8 7	156,892 4 4	162,337 9 0
Mails and Express	20,304 12 11	24,215 14 2	23,438 17 9	31,678 3 0	30,172 3 10
Baggage	53,907 8 11	55,734 14 3	56,365 11 3	59,449 14 8	62,592 1 6
Freight Local	2,562 9 9	1,946 19 5	2,113 19 9	2,127 19 5	2,206 19 0
" Through	625,810 1 1	634,263 16 9	669,207 19 9	713,047 13 9	716,048 2 3
Live Stock	128,685 17 9	135,237 13 10	120,470 10 3	117,610 0 11	167,493 0 11
Rents and Sundries	56,410 15 7	32,691 18 11	46,781 13 6	66,347 10 0	72,951 16 1
	8,118 13 10	6,970 9 9	5,091 3 5	·4,487 16 2	4,139 1 7
Gross Receipts					
Less Cartage and Refunded Fares	1,378,380 08 4	1,358,905 10 3	1,378,282 7 3	1,457,524 12 2	1,529,257 9 8
	26,174 9 5	30,413 15 0	27,106 2 2	26,739 14 2	33,496 19 10
Total					
	1,352,205 18 11	1,328,491 15 3	1,351,176 5 1	1,430,784 18 0	1,495,760 9 10
EXPENDITURE.					
Maintenance of Way Stations, &c.	129,160 1 10	130,097 10 10	136,647 18 2	134,833 11 0	135,215 16 10
Renewals of Way, &c. &c	168,919 17 9	137,638 15 8	*163,496 5 1	203,811 3 7	140,526 8 3
Working and Repairs of Engines	270,211 5 1	298,299 0 4	320,816 10 4	330,934 8 2	340,608 17 10
Repairs and Renewals of Cars	83,292 7 1	86,278 10 3	92,139 9 11	99,989 10 11	112,937 4 1
Traffic Expenses	258,296 11 2	264,514 1 1	266,793 19 11	268,048 17 7	295,643 5 5
General Charges	29,368 14 2	31,121 15 11	30,568 19 8	29,907 18 8	30,982 19 10
Miscellaneous Expenses	18,467 19 4	18,322 19 10	14,567 18 0	23,787 9 9	23,103 1 2
Rent of Rolling Stock	4,523 6 6	11,019 0 9	6,424 14 0	9,693 19 3	28,421 6 3
Tolls paid other Co's. and use of Stations	5,772 18 6	3,748 6 2	5,823 9 4	8,250 0 1	10,578 14 4
Total Working Expenses and Renewals	968,013 1 5	981,040 0 10	1,037,279 4 5	1,109,256 19 0	1,018,018 4 0

Net Revenue	384,193	17	6	347,451	14	5	303,307	0	0	321,527	19	0	377,760	5	10
Less Loss on American Currency	73,113	9	6	83,383	15	7	65,267	7	9	63,159	1	1	46,085	18	5
Leaving the balance of	311,080	8	0	264,068	16	10	238,039	13	6	258,368	17	11	331,674	7	5
Add Balance from last year	80	2	0	200	12	6	†17,677	2	3	8,934	14	9	20,673	5	8
" Second Pref. Int. account	36	19	8												
Am't applicable for the following pay'mts	311,197	9	8	264,269	11	4	220,362	11	3	267,303	12	8	*352,348	13	1
Amount due Postal Bondholders	43,640	8	2	33,873	16	7	37,081	17	0	32,578	13	4	28,491	4	2
Interest paid on Loans	4,646	1	11	3,458	5	9	3,973	19	0	3,904	0	11	8,032	9	5
" Mortgage to Bank of U. C.	8,847	12	2	8,847	12	4	8,847	12	4	8,667	12	4	8,847	12	4
Interest, Loans, Bankers' Balances, European Exchange, &c.	18,914	13	9	6,457	3	4	6,974	8	0	16,106	11	0	8,577	15	0
Interest, British American Land Co. Deb.	1,232	17	6	1,232	17	6	1,232	17	6	1,232	17	6	1,232	17	6
" Montreal Seminary Debentures	1,232	17	6	1,232	17	6	1,232	17	6	1,232	17	6	1,232	17	6
" Island Pond "	5,400	0	0	5,400	0	0	5,400	0	0	5,400	0	0	5,400	0	0
" Portland Sinking Fund	4,058	5	8	4,058	5	8	4,058	5	8	4,936	0	0	4,739	14	11
Atlantic & St. Lawrence Lease	63,371	11	2	63,872	2	0	64,177	17	10	65,010	19	5	67,081	18	11
Detroit Line Lease	23,500	0	0	23,500	0	0	23,500	0	0	23,500	0	0	23,500	0	0
Montreal & Champlain Co.	20,547	19	0	17,979	17	4	17,585	2	2	16,219	18	0	22,402	4	0
Buffalo & Lake Huron Co.	42,510	2	11	34,145	9	8	34,229	19	6	16,387	11	8	44,250	0	0
Equipment Bond Int.	13,016	0	0	31,858	0	9	21,556	0	0	21,556	0	0	22,556	0	0
First Preference Bond Int.	58,403	8	3	36,264	11	11							67,787	1	0
Balance carried next year	200	12	6	*9,312	13	9	8,934	14	9	44,420	12	8	16,387	18	11
				+17,677	2	3									
	311,197	9	8	264,269	11	4	220,362	11	3	267,303	12	8	352,348	13	1

* Less interest received on Bonds in hand. † Debit balance deducted.

The Grand Trunk Line extends from Portland, Maine, to Detroit, Michigan, a distance of 854 miles, to which adding branch lines owned, leased and operated, 523, gives the total length of the Grand Trunk system 1,377 miles. Gauge of track, 5ft 6 inches. Weight of rails 60 to 75 lbs to the yard. The line is divided into operating districts, thus :—

Detroit (Detroit to Port Huron)	59	miles
Western, (Sarnia to Toronto)	168	do
Central, (Toronto to Montreal)	333	do
Buffalo and Goderich and Branch	168	do
Province Line Division	40	do
Riviere Du Loup Branch	126	do
Champlain, (Montreal to Rouse's Point)	49	do
Montreal and Island Pond	145	do
Quebec and Richmond	96	do
Three Rivers Division	35	do
Portland, (Island Pond to Portland)	150	do
Lachine Branch	8	do
Total	1,377	miles

The following railways are worked under lease, and are included in the above.

Atlantic and St. Lawrence, Portland, Maine, to Island Pond, Vt.	150	miles.
Chicago, Detroit and Can. Grd Junction R. R., Detroit to Port Huron	59	do
Montreal and Champlain, Montreal to Rouse's Point, N. Y.	49	do
Buffalo and Lake Huron, Fort Erie to Goderich	161	do
Total leased lines	419	miles.

ENGLISH DIRECTORS.—Richard Potter, Gloucester, Eng., President. Thomas Baring, M.P., London, Eng., E. C. ; Robert Gillespie, Gravenhurst ; Kirkman Daniel Hodgson, London, E. C. ; Grosvenor Hodgkinson, M. P., Newark ; Graham Menzies, London, W. ; John Swift, London, W. ; Captain Tyler, Hampton Court ; Lord Wolverton, London, E. C.

CANADIAN DIRECTORS.—C. J. Brydges, Montreal ; Hon. James Ferrier, Montreal ; Wm. Molson, Montreal.

MANAGING DIRECTOR.—C. J. Brydges, Montreal.

SECRETARY AND TREASURER.—Joseph Hickson, Montreal.

AUDITORS.—William Newmarch, London ; J. G. T. Child, Manchester

GREAT WESTERN RAILWAY.

On the 6th March, 1834, an Act was passed by the Canadian Legislature to incorporate the London and Gore Railroad Company. Among the incorporators were Allan Napier (afterwards Sir Allan) McNab, George J. Goodhue, Edward Allan Talbot and seventy others, a number of whom were prominent public men in those days. Power was taken in the charter to construct a "single or double track, wooden or iron railroad," from London to Burlington Bay, and also to the "navigable waters" of the river Thames and Lake Huron, and "to employ thereon either the force of steam or *the power of animals*, or any mechanical or other power." The capital was fixed at $400,000 (£100,000), in 8,000 shares of $50 each; and in the event of the continuation to Lake Huron, the capital might be doubled. The time for the completion of the road was limited to twelve years.

Nothing was done under the powers granted by this Act. In 1846 when it was about to lapse, an Act was passed reviving the Act of 1834, with amendments. One of these amendments was to change the name to "The Great Western Railway Company." Power was taken to build the line to some point on the Niagara River; the capital was increased to $6,000,000 in 60,000 shares of $100 each; and the time allowed for the completion of the line was extended to 20 years.

Of the capital so authorized 55,000 shares were promptly subscribed in England, and only 5,000 shares in Canada. This led to the passage of an Act in the following year (1846), "for the purpose of affording just and proper protection to the English shareholders." This Act provided for the appointment of a Committee not to exceed eleven persons, residents of London, England, with very large powers of regulating the management of the Company's affairs. In 1849 this Act was repealed, and British and Canadian shareholders were placed on the same footing; the number of directors was increased from seven to eleven.

The main line leaves the Niagara river at an elevation of 326 feet above Lake Ontario. It gradually descends to the level of the lake at Hamilton, where grain and general freight warehouses are erected on the wharf. The line then steadily rises till the summit level is reached, 83 miles west of the Suspension Bridge, where the elevation above Lake Ontario is 762 feet. From thence it again gradually falls till it reaches the Detroit river at Windsor. The steepest grade is that ascending to the west from Hamilton, averaging 50 feet per mile for 10 miles. From Komoka westward, for 100 miles, the line is nearly level, and there are 57 miles of this length in a single straight line.

The spirit of speculation which prevailed from 1853 to 1856 was a source of embarrassment and expense to this and every other Company constructing lines in the Province. This state of things was to be attributed chiefly to the railways. So great was the demand for labor, live stock, timber and materials of all kinds by the competition which existed, that prices increased 30, 40 and 50 per cent. Contractors who had undertaken to build sections of this Railway at low estimates failed, one after another, and the works had to be relet at advanced figures. As in the case of nearly all the railways the original estimates fell far short of the actual cost. It was found in 1854 that an estimate made by the Company's engineer in 1852 for the main line was about a million and a half of dollars under the mark. A single instance will show that regardless of the advance in materials these estimates were ill-devised, and little else than mere guess work ; the rails were estimated at their first cost in Wales, with no allowance for freight, insurance or duties. The cost of land was put down in the estimates at $60,000, whereas the amount actually expended under this head was $700,000.

The share capital was raised under the authority of five different acts of the Legislature. The date of these acts with the amount of capital authorized to be raised are as follows :—

	Shares.	Amount.
8 Vic. cap. 86, of March 29, 1845	60,000	$6,000,000
16 Vic. cap. 99, of April 22, 1853	20,000	2,000,000
18 and 19 Vic. cap. 176, of May 19, 1855	60,000	6,000,000
16 Vic. cap. 44, of Nov. 10, 1852, (H. & T. Act)	18,000	1,800,000
16 Vic. cap. 101, of Ap. 22, 1853, (Sarnia Act)	20,000	2,000,000
	178,000	$17,800,000
G. W. Amend't Act 22 Vic. cap. 116, of 16th Aug. 1858		8,000,000
Total capital..................		$25,800,000

The sum of $3,850,000 (£770,000 stg) was advanced by the Government under the provisions of the Main Trunk Guarantee Act. It was provided that this loan was to pay 6 per cent interest, and that 3 per cent was to be annually set apart as a sinking fund. This large amount of public money was not hopelessly sunk as in the case of the advances of the Grand Trunk and Northern ; large sums have been repaid and the whole is now in such a shape that its liquidation is rendered certain.

The existing hostility between this Company and the Southern Railway project is of old standing. In 1857 the Directors of the Great Western report that "during the last twelve months considerable discussion has arisen in regard to the projected Southern Line through Canada, which was last summer attempted to be forced upon this Company. In the last session of the Provincial Legislature many disgraceful disclosures were made as to the past history of that scheme arising out of the rival claims of certain parties, to the control of the line. These disclosures, showing an extent of bribery and dishonesty which have been rarely paralleled in the history of any joint stock undertaking, and which called forth the marked and emphatic denunciations of committees of the Provincial Parliament cannot fail to increase the satisfaction of the shareholders that this Company was preserved from any connection with such a scheme."

In October, 1857, the Directors were authorised to advance the sum of $750,000 to the Detroit and Milwaukee Railway, to help that line out of certain difficulties into which it had fallen. The Directors in reporting in favor of this advance say that they had "caused a careful examination to be made into the statements furnished by that Company as to its affairs and accounts and the result of a complete and thorough investigation showed that the sum of $750,000 would be sufficient to meet the claims of the secured creditors, and leave enough to open the line and provide rolling stock." The loan was accordingly made, secured by a mortgage in favor of Mr. C. J. Brydges, T. Reynolds and H. C. R. Becher, three of the Canadian Directors. Under the conditions of this mortgage the entire control of the affairs of the Detroit and Milwaukee was placed in the hands of directors to be nominated from time to time by the Great Western Company. Arrangements were then made for the completion of the Detroit and Milwaukee Line to Grand Rapids and through to Lake Michigan. It was opened for traffic through in September 1858. Most favorable results to the revenue of the Great Western were expected to follow from this transaction. The case was stated thus to the shareholders of the Great Western Company in the Directors' report of March, 1859.

"The amount of gross traffic required to pay the interest upon the whole borrowed capital of the Detroit and Milwaukee Company ($4,956,000 or about £1,000,000 stg. on a line of 185 miles in length) allowing 50 per cent for working expenses, is only about $76 per mile a week, which places the success of the Company beyond any doubt; and for whatever money is still required for that Company there is the certainty of the traffic being quite sufficient to yield a full security and a handsome return." A further sum of $600,000 was loaned on the same terms as the first loan. The cost of the Detroit and Milwaukee line with a fair amount of rolling stock was about $40,000 per mile.

Proceedings were taken in 1860 to foreclose two mortgages held by the Great Western on the Detroit and Milwaukee Line, and on the 16th of April of that year Mr. C. J. Brydges was appointed Receiver. These proceedings were taken with a view to protect the interests of the Great Western, and to provide for the efficient working of the Line till some desirable arrangements could be made. On the 6th of August, however, a decree of sale was granted by the court of Chancery of Michigan, and on the 4th October the Railway was put up for sale at Detroit and purchased by Messrs. Gray & Reynolds for the nominal sum of $1,000,000, as joint trustees for a new company to be formed, on certain conditions to be carried out before June, 1861.

After consultation had between the Directors and Mr. Brydges, who went to England for the purpose, it was decided to settle the difficulty in the following way :—1st, "That all the coupons on the first and second mortgage bonds o the Detroit and Milwaukee Company, up to and including 15th May, 1862, be funded and exchanged for bonds bearing 7 per cent interest, maturing on 1st January 1866, to enable the Company to use the net earnings of the Line for paying certain of the debts of the Company, ordered by the Court to be paid in cash. The interest on these deferred bonds will be paid half-yearly in cash. 2nd, That the amount of the Great Western Company's foreclosed mortgages, with interest up to the date of the sale of the road, 24th October, 1860, be converted into Preference shares, to be secured by a third mortgage (subject only to the priority of the first and second original mortgages).on the entire Line, to carry interest up to 7 per cent., per annum, out of the net earnings of the Line. 3rd, That a fourth mortgage, subject to the above, be created, to secure a further issue of Preference shares, also bearing interest up to 7 per cent., per annum, to an amount equal to the floating debt of the Company—the holders of which shall be entitled to receive such stock at par in payment of their claims. 4th, That, under the circumstances of the case, after the payment of the sums ordered by the Court to be paid in cash, the interest on the two above mentioned classes of Preference shares up to 31st December, 1864, be paid equally *pro rata* out of the net earnings of the Line, after paying interest on the two prior mortgages. After 31st Dec., 1864, interest upon the different classes of securities to be paid in the order of their legal priorities. 5th, That the ordinary shareholders shall receive ordinary shares in the new Company to the extent of 20 per cent of the par value of the old shares."

These proposals on the part of the Great Western did not meet with the unanimous acceptance of the creditors ; and a suit was commenced by the Commercial Bank of Canada against the Detroit and Milwaukee Railway (to this suit the Great Western was made a party) to recover the sum ot £250,000, being an advance made by the bank to the Detroit Company. The case was tried at Kingston before a jury and afterwards at Toronto, and being decided in each instance against the Great Western it was taken to the Court of Error and Appeal, and ultimately to the Privy Council. Meantime the overdue interest on the loan of £250,000 increased to more than half the amount of the principal, being in September 1863, £150,000. The decision of the Judicial Committee of the Privy Council was delivered on the 27 July, confirming that of the Canadian Court of Error and Appeal and dismissing the appeal with costs against the Bank.

An arrangement was finally arrived at, the main principle of which is that all claims against the Great Western Company were withdrawn, the Detroit and Milwaukee Company agreeing to set aside a moderate annual sinking fund to liquidate the claim of the Bank, and in the meantime to issue its bonds in satisfaction of this and other claims. These bonds participated to a limited extent in the surplus earnings of the Detroit and Milwaukee Road *pari passu* with the original loan made by the Great Western and the interest accrued thereon. An additional issue of securities to the Great Western Company to the amount of $393,000 was made by the Detroit and Milwaukee on account of arrears of interest, making the total amount of loan and accrued interest $2,100,000.

Though the anticipations of the Directors respecting the Detroit Line were not realized, and though the interest on the loan was not paid yet considerable increase of traffic resulted. This loan proved to be a serious affair for the Great Western, and so important was its influence on the Company's position regarded that in the report of April, 1860, this passage appears. "The critical financial position of the Detroit and Milwaukee Company—its close connection with and indebtedness to this undertaking—the continued depression of the receipts of the latter—constitute apparently a state of things so unsatisfactory that it is desirable that the fullest information should be afforded, and that the opinion of perfectly impartial persons should be taken upon the position of the concern. The Directors therefore recommend that a committee of proprietors be appointed at the approaching meeting for the purpose of investigating the affairs of the Great Western Railway Company with all requisite authority, and that as soon as their report is prepared a special meeting shall be summoned for its consideration."

In 1858 the Great Western in common with other lines suffered from a serious falling off in its traffic. It was less as compared with 1857 by 23 per cent., and that of leading American lines showed a decrease ranging from 14 to 25 per cent. In the half year ending July, 1859, no dividend at all was paid. The disheartening position of affairs at that time was described in this frank and truthful language by the Directors. "In placing this statement," (for the half year), "before the shareholders, the Directors cannot but express their extreme concern and disappointment at the altered position which it exhibits of the Company's affairs. In the report placed before the meeting of the 6th April last, a sanguine hope was entertained that the worst was then over, and that a gradual improvement from the state of depression the Company was at that period laboring under, might fairly be calculated upon. Unfortunately this has not been borne out by the result, and this company has had to sustain, during the last half year, a continuance of the most adverse circumstances in common with every other railway on the Northern portion of the American continent. The traffic of the line both through and local has undergone a diminution during the last three years, of which we have no parallel in the history of railways in this country, and though the exertions of the Executive in Canada have effected a most important reduction in the working expenses, this has not been adequate to sustain the Company's position and earn a dividend."

The earnings for the first half of four successive years fell off in the remarkable manner shewn by these figures :—

Earnings of first half of 1856	$1,160,502
Earnings first half of 1857	1,065,720
Earnings first half of 1858	854,693
Earnings first half of 1860	736,994

showing a falling off of $443,688 as between 1856 and 1859, while an increase was naturally to be expected. Were it not for an important diminution in the Company's expenses at the same time the effect upon its financial position must have been serious. The next dividend was also foregone.

At the same time that the Company's traffic, both in freight and passengers fell, off so steadily and rapidly, a new obstacle stared the Directors in the face. The Engineer, Mr. Geo. Lowe Reid, reported that during the half year, commencing February 1861, a "renewal of the rails of the whole Main Line and of the Toronto and Galt branches will have to be systematically begun." He estimated that this renewal of rails would have to be completed within five years. As there were 250 miles to be relaid in the five years, an annual average of 50 miles of rails had to be put down. He estimated that the sleepers, which were rapidly giving out, would all have to be replaced within three years from 1st Feb. '61, requiring an average of 160,000 sleepers per annum.

The cost of these renewals of the permanent way including new joint fastenings, and the labor of relaying the rails and sleepers, &c., was stated at $285-000 each year for the five years. The rails had only been six and a half years in use, and their average life would not exceed eight years. This very unsatisfactory result arose from the inferior quality of the iron in the case of the fish rails, and from the defective form of the rail and its joint fastening, combined with a poor quality of metal, in the case of the bridge rails.

The Engineer also reported that the wooden bridges, amounting to 13,915 lineal feet, on the Main Line and Galt Branch would all have to be rebuilt within the five years before referred to. The cost of renewing these entirely in timber is stated at $230,000, spread over a period of five years. Owing to the fact that these wooden structures never last in this country more than ten to twelve years, the Engineer very properly recommended that iron and stone be largely used in the new bridges.

Mr. Reid estimated the total annual expenditure for the renewal of the permanent way, including bridges and fences on the Main Line and Galt Branch as follows :—

1st year	commencing	Feby.	1861	$237,000
2nd	do.	do.	Feby. 1862	272,000
3rd	do.	do.	Feby. 1863	315,000
4th	do.	do.	Feby. 1864	410,000
5th	do.	do.	Feby. 1865	538,500
		Total		$1,772,500

These expenditures were estimated to be in addition to the ordinary repairs or maintenance of way, which was then at the rate of $142,000 a year.

In common with all our leading railways the Great Western suffered severely from bad rails. The original track consisted of 38½ miles of compound rails weighing 66 and 80 lbs. per lineal yard ; 156 miles of the U or bridge rail of 66 lbs. to the yard, and 34½ miles of the fish-jointed rail of 65 lbs. to the yard.

By the end of July, 1860 the track was so altered as to consist of 116 miles of fish-jointed rails, 66 lbs to the yard, and 112 miles of the U rail, 65 lbs to the yard: showing that in the 6½ years the whole of the compound rails, and 63 miles of the U rails had been replaced by fish-jointed rails. The Toronto Branch was laid with fish-jointed rails throughout. This kind of rail proved to be very inferior in quality, especially those laid down on the Toronto Branch. On many sections of the line where there were sharp curves or heavy gradients they did not last two years, and their average was as low as six years. They were made from soft iron of poor quality and were welded badly, and in consequence laminated to an unexampled extent even under ordinary traffic. The U rails were made from harder iron, but were of a form badly adapted to our climate, and being supported on cross sleepers with merely a flat plate at the joint to which the ends of the rails were bolted or spiked, the track was wanting in vertical stiffness. In alternating frosts and rains or thaws the road was sure to become uneven, no matter how solid the road bed; and in consequence these U rails, being perforated in the lower flanges with bolt holes, would break to an alarming extent. Mr. Reid states that sometimes in one day of intense frost as many as 20 rails were broken, some of them in two places, by a passing train.

It is manifest from the general experience with English rails, that those made in the early days of railroads were much superior in quality to those manufactured since 1850; as the demand from abroad increased and the trade expanded competition became keener, prices consequently diminished, and the quality of the iron rapidly degenerated, till, as an American railway authority states, many lots which were sent to the United States were not worth the expense of laying down. One lot purchased in Newcastle lasted only a trifle over four years.

The great expense of re-rolling rails here, being about $30 per ton for re-rolling and supplying the loss in weight, as against about one-fourth of that sum in England—induced the Company to take steps for establishing a rolling mill for their own use. The rolling mill at Hamilton was therefore commenced in 1862 or '63 and completed in the early part of 1864. It cost about $167,500. The size of the mill is 120 x 135 feet, employs about 130 men, and working day and night has a capacity of 7,000 tons (70 miles of track) per year.

In 1869 an arrangement was made with the Government of the Dominion respecting the extinguishment of the Company's indebtedness to the Government. It was agreed that the principal with accrued interest to the 1st January 1869, should be commuted for the sum of £668,515 7s. 0d., payable in annual instalments, the unliquidated balance, year by year, to bear interest at the rate of 4 per cent., per annum, instead of 6 per cent as before. This was regarded by the Directors as being equal to a reduction in the debt of £180,000. One of the conditions of the bargain was a payment in cash of £100,000 on the 1st Feby 1869, which was complied with. For the purpose of raising the necessary money to carry out this arrangement, an issue of preferred stock was made to the amount of $5,000,000 bearing interest at 5 per cent at the rate of 80 per cent of its nominal value.

On the 12th June, 1867, an agreement was made with the Grand Trunk Railway providing that equal fares and rates should be charged from all competitive points; the gross receipts of each Company for local passenger and freight traffic between certain competitive points, and also between these places and competitive points of the lines to the east, to be divided in such portions as agreed upon; the Grand Trunk to be permitted to send their loaded cars, from any station on their lines of railway east of Toronto to any station on the lines of the Great Western west of Toronto, such cars being handed over to the Great Western at Toronto; these cars to be returned loaded with freight from stations on the Great Western line to any station on the Grand Trunk, east of Toronto; in like manner the G. W. R. are permitted to send their loaded cars from any station on their railways, except Toronto, to any station upon the section of the G. T. R. line west of and including the Buffalo and Goderich line, such cars to be handed over to the G. T. R. at Paris, the same cars to be returned loaded with freight for stations on the G. W. R. line, Toronto excepted. The rates charged from local stations on the G. W. R. to Toronto for places east thereof on the G. T. R. are the same as those charged by the G. W. R. from the same points to Suspension Bridge, and when this would not apply equal rates per ton per mile was to be charged. The rate to be charged to and from local competing stations west of Toronto to and from the stations of Toronto and Hamilton are the same whether carried over the G. T. R. or the G. W. R. The through rate to be charged from Montreal to points on the G. T. R. and the G. W. R. lines west of Toronto and Hamilton, and *vice versa*, shall be such as agreed upon. Teaming freight to and from competing places to be abolished. Passenger trains of both companies to be so timed as to connect at Toronto and Paris. The rates for all through traffic to be such as are agreed upon between the managers of the two lines. Any projected competing lines west of Toronto to be either undertaken and constructed jointly by the two Companies, or the option to be given by the one to the other Company to occupy and work the same jointly upon such terms and conditions as may be agreed upon. Some other clauses are contained in the agreement respecting the mode of settling of disputes, &c., and it is provided that the agreement shall remain in force seven years from the 1st August 1867, unless sooner terminated, which either may do on giving six months notice to the other Company.

In 1869 an arrangement was entered into between the Great Western, the Michigan Central, and the Detroit and Milwaukee Railways, for the period of two years, relating to their through traffic. By this arrangement the receipts from through traffic are to be divided between the three lines in the proporportion of 48 per cent. to the Great Western, 48 per cent. to the Michigan Central, and 4 per cent. to the Detroit and Milwaukee. The length of the three lines is Western 239 miles, Michigan Central 229 miles, Detroit and Milwaukee 189 miles. The proportion of earnings for the purposes of the agreement was based on the results of the two previous years' through traffic in the case of the two first named lines, and on one years' traffic of the Detroit and Milwaukee Line.

At Suspension Bridge, the connection is formed with the New York Central Railroad on the American side by means of the Suspension Bridge, which was opened for trains in March, 1855. At Windsor, the connection is formed with the Michigan Central and Detroit and Milwaukee Railroads by means of ferry steamers, the width of the river being half a mile. One is an iron double-ender steamboat, 240 feet in length, which takes over a whole passenger train on its two tracks, or 14 freight cars. The other is a large wooden steamer with a spacious saloon on deck, on which passengers only are transferred.

An Act was obtained from the Dominion Legislature two sessions ago repealing so much of the Act of 1851, as required the Company to construct the railway and branches with a gauge of 5 feet 6 inches, and authorizing the alteration of the gauge to that of 4 feet 8½ inches, commonly called the narrow gauge. Since this power was confirmed the greatest energy has been displayed in making the change, and now the whole line from Windsor to Komoka, and from Hamilton to Toronto and Suspension Bridge, (153 miles) the gauge is four feet 8½ inches. The remainder of the line is a mixed gauge of 5 feet 6 inches and 4 feet 8½ inches, and the remaining branches are 5 feet 6 inch gauge.

Like the others of our three leading railway companies the *Great Western* tried the experiment of running lake steamers in connection with their line; but with anything else than gratifying results. The Directors complain bitterly of the opposition of the Huron and Ontario Railway steamers, attributing their want of success in 1863 chiefly to this cause.

The "Canada" and the "America" were built by the Company to run upon Lake Ontario between Hamilton and Oswego and were placed on that route on the 28th June, 1855. They cost $350,669. From these steamers important advantages were expected, but they proved to be a source of loss to the extent of nearly $60,000. In 1856 it was arranged to place them on a new route forming a daily line between Hamilton and Cape Vincent, Brockville, Prescott and Ogdensburgh; but this scheme resulted in another loss of $23,000, and on the 23rd August the boats were withdrawn for the season. They were finally sold to parties connected with the Detroit and Milwaukee Railway and payment accepted in shares of that line bearing 7 per cent interest.

DETROIT TUNNEL.

By a charter obtained from the United States Congress and by an Act passed in May, 1870, powers were given to the Michigan Central Railway, and the Great Western Railway to construct a tunnel under the Detroit river for railway purposes. The capital stock is fixed at $6,000,000 in shares of $100 each. Directors may be elected when $1,000,000 are subscribed; and that amount must be paid in within two years and the work commenced and the Tunnel completed in six years. The Boards of the two Companies have decided to go on with the scheme, and the preliminary works have been commenced. They have agreed to expend $60,000 between the two Companies.

The principal dimensions of this great work will be :

Length between Michigan Cen. R. R. Station and Great Western Station	2.84 miles.
Length across River	3,000 feet.
Length from Portal to Portal	8,568 "
Maximum gradient	1 in 50 "
Depth under River	97 "
Depth of water over Tunnel	51 "
Depth of clay above Tunnel	12 to 20 "
Diameter of Tunnel	18½ "
Estimated Cost	$3,000,000

BRANCH LINES.

The following branches are worked by the Great Western Company. 1. The Erie and Niagara Railway—Fort Erie, opposite Buffalo, to Niagara—31 miles. 2. The Toronto Branch—Hamilton to Toronto, where a connection is formed with the Grand Trunk Railway—29½ miles. 3. The Galt and Guelph branch—Harrisburgh to Guelph—27½ miles. 4. The Sarnia branch—Komoka, west of London, to Sarnia, opposite Port Huron—51 miles. 5. A line from the Sarnia Branch at Wyoming, to the oil region of Petrolia—5½ miles.

GALT AND GUELPH BRANCH.

In 1852 Isaac Buchanan and 17 others were incorporated as "the Galt and Guelph Railway Company." The capital was limited to $560,000, in shares of $100 each, and power was granted to borrow the sum of $200,000. An arrangement was entered into with the Great Western Company by which that Company was to supply the Galt and Guelph Company with the rails required for this line, receiving from them first mortgage bonds of the Galt and Guelph Railway for the value of the rails. The Line was completed to the town of Preston, 4 miles from Galt, and opened for traffic on the 28th November 1855, Difficulty was encountered at this stage ; the town of Guelph came forward with a subscription of $80,000 to be paid in cash to the Galt and Guelph Company. The Galt and Guelph Company then undertook to issue additional first mortgage bonds to a sufficient amount to complete the road. It was included in the bargain between the two Companies that the Great Western was to work the Line at cost, and after deducting the interest at the rate of 6 per cent. on the bonds issued to the Great Western Company, to refund the balance to the Galt and Guelph. The total bonds so issued, including those for iron, were $260,000. Mr. Reid's estimate for that portion of the Galt and Guelph Line, from Preston to Guelph, 11½ miles, exclusive of the rolling stock, was $267,200, which was something less than the actual cost. The contract was let to A. P. Macdonald & Co., in March 1856, and was completed by the end of 1857, and opened for traffic 11th September. It is a substantially constructed line. The total expenditure on this road up to March 1858, including iron, &c., was $440,169.

This branch did not prove profitable, and within three years after it was opened the property became hopelessly embarrassed. In 1860 the Directors of the Great Western reported that this Line "not having earned or paid any interest upon the amount expended on it by this Company the mortgage taken for our advances has been foreclosed." It then became the property of the Great Western Company as mortgagees, for the sum of $804,788 80.

TORONTO AND HAMILTON BRANCH.

In 1852 an Act was passed incorporating 26 gentlemen as "the Hamilton and Toronto Railway Company" with power to raise a capital of $1,000,000 in shares of $100 each, and to build a line from Hamilton to Toronto, 38 miles in length. The contract was let to Mr. George Wythes. In the next year arrangements were made for the lease of the Line to the Great Western Company, at a rent of 6 per cent on its cost, together with an equal participation in any dividends earned by the Great Western beyond that amount. The Great Western supplied the rolling stock and station buildings at a cost of about $400,000. The Line was open for traffic on the 3rd December 1855. An arrangement for amalgamation was made with the Great Western in 1856, which went into effect in that year.

In the Great Western accounts for July 1856, this branch is debited with an expenditure of $1,860,556, the cost of the Line and equipment.

SARNIA BRANCH.

In 1853 the London and Port Sarnia Railway Company was incorporated with a capital of $2,000,000, and consisted mostly of the same gentlemen constituting the Galt and Guelph Railway Company. This branch is 51 miles in length. Power was taken to amalgamate with the Great Western Company. The contract was let for $1,440,000, but the work was suspended at the instance of the Company in 1854, under an agreement with the Grand Trunk Company to that effect. In the early part of 1856 it was recommenced under the terms of the original contract. The total cost of the branch, including rolling stock, was estimated at $1,800,000 to $2,000,000. This branch was finally opened for traffic on the 27th December, 1858. The total cost for lands, works, bridges, permanent way, stations, warehouses, and all incidental charges to 31st January 1862, was $1,873,666.

THE "CANADA AIR LINE RAILWAY."

Is a loop line under construction of 146 miles, from Glencoe to the City of Buffalo. The road will leave Glencoe, a station on the Great Western main line 89 miles from the Western terminus, and will proceed, with but little deviation from a straight line to Fort Erie on the Niagara river, directly opposite to the City of Buffalo, an unbroken connection with the various American railroads centring in that city being made by the International bridge now in course of construction. The Act authorising the loop line confers running powers over 44 miles of the Buffalo and Lake Huron branch of the Grand Trunk railway, from a station called Canfield to Fort Erie, if terms can be agreed upon. The engineer's estimate of the cost of constructing the road is $25,000 per mile, which includes steel rails, bridges, station buildings, and approaches, and land; a further supply of rolling stock will not exceed $5,000 per mile in addition, and the engineer undertakes that, for this outlay, the permanent way shall be equal to that of the New York Central and Hudson River railroads. It was long foreseen that the necessity would arise either to build this loop as a relief to the main line, or to double the existing main track. The constantly increasing passenger and freight traffic carried over the narrow gauge route in connection with the continued extension of the American railroads westward, even up to the Pacific Ocean, has nearly reached the capability of a single track of rails, and is already equal to the tonnage carried over many double track railways in this country. In order to facilitate this traffic the Company have taken up the broad gauge line of rails on the main line as well as branches so as to work the whole traffic over the ordinary narrow gauge of the American roads. The engineer estimated the cost of doubling the present main line from Suspension Bridge to London at about $25,000 per mile. Preference was therefore given to the construction of a loop line, which not only makes a shorter through route, but traverses a new district of country, the local traffic on which affords a net revenue equal to 5 per cent. per annum on its cost. The route adopted along the flat table land, level with Lake Erie, affords easy gradients as compared with the existing main line, so that the haulage of heavy through freight trains, and fast passenger trains, will be greatly facilitated. Considerable economy will thus be effected in the maintenance of way and in carrying greater loads with the same engine power; these two items alone are calculated to effect a saving in money value of $125,000 per annum as applied to the same tonnage carried over the present main line. A very great advantage to be expected from this loop line, is the alternative route it will open to New York.

PETROLIA BRANCH.

This branch was formally opened for traffic on the 17th December 1866. It cost £10,551 14s. to July 1867, for five miles of railway including rails, station buildings, &c., and the traffic earnings of the first six months were £8,451 10s.

WELLINGTON, GREY AND BRUCE.

This Line is substantially an extension of the Galt and Guelph railway northward, and is open to Alma. An agreement exists between this Company and the Great Western, by which the latter have agreed to supply the rolling stock and work the road at 70 per cent. of the gross earnings. An account is to be kept of the railway traffic exchanged between the Great Western and this Line, and 20 per cent of this traffic shall be set aside annually and appropriated to reduce the capital cost of the Line, so that in the course of years the branch will gradually become a part of the Great Western system.

GREAT WESTERN RAILWAY.

ROLLING STOCK, &c.	Fiscal Year ending 31st January					
	[illegible]	[illegible]	[illegible]	[illegible]	[illegible]	[illegible]
Locomotives........No.	[illegible]	60	[illegible]	[illegible]	[illegible]	[illegible]
Passenger Cars ... "	[illegible]	114	[illegible]	[illegible]	[illegible]	[illegible]
Baggage, Post Office, and Express Cars "	[illegible]	41	44	[illegible]	[illegible]	[illegible]
Freight Cars .... "	1,[illegible]	1,2[illegible]	1,[illegible]	1,[illegible]	1,[illegible]	[illegible]
Gravel Cars ... ... "	[illegible]	[illegible]	[illegible]	[illegible]	[illegible]	[illegible]
Mileage of Locomotives	[illegible]	1,[illegible]	1,[illegible]	1,[illegible]	1,[illegible]	[illegible]
" Pass. trains	[illegible],519	617,[illegible]	677,[illegible]	[illegible]	[illegible]	[illegible]
" Freight "	[illegible]	614,[illegible]	[illegible]	[illegible]	[illegible],14[illegible]	[illegible]
Passengers carried, No.	[illegible],411	[illegible]	[illegible]	[illegible]	[illegible]	[illegible]
Freight carried, No. tons	[illegible]	176,[illegible]	[illegible]	19[illegible],70[illegible]	177,[illegible]	[illegible]
REVENUE.	£ Stg.	£ Stg.	£ Stg.	£ Stg.	£ Stg.	£ Stg.
Passengers	[illegible]	[illegible],41[illegible]	[illegible]	[illegible]	[illegible]	[illegible]
Freight and Live Stock,	[illegible]	[illegible]	[illegible]	[illegible]	[illegible]	[illegible]
Mail and Express... ..	12,[illegible]	16,[illegible]	20 [illegible]	21,[illegible]	[illegible]	[illegible]
Sundries	[illegible]	1,[illegible]	2,11[illegible]	2,[illegible]	2,[illegible]	2,[illegible]
Total Receipts, ..	[illegible]	677,[illegible]	[illegible]	[illegible]	[illegible]	[illegible],718
EXPENDITURE.						
Maintenance of way, works and stations..	[illegible],317	[illegible]	[illegible]	[illegible]	[illegible]	[illegible]
Locomotive power	[illegible]	[illegible]	[illegible]	[illegible]	[illegible]	[illegible]
Repairs and renewals of Cars	11,[illegible]	18,[illegible]	[illegible],518	17,[illegible]	[illegible]	14,[illegible]
Transit expenses —Passengers and Freight..	[illegible]	[illegible]	11[illegible]	[illegible]	[illegible]	[illegible]
General Charges	[illegible]	14,[illegible]	[illegible]	[illegible]	14,[illegible]	[illegible]
Total working exp's	[illegible]	[illegible]	[illegible]	[illegible],71[illegible]	[illegible]	[illegible]
Taxes, Insurance, &c. ..	2,[illegible]	3,[illegible]	4,[illegible]	3,[illegible]	4,[illegible]	[illegible]
Susp. Bridge Rent	......	6,[illegible]	9,[illegible]	9,[illegible]	9,[illegible]	[illegible]
Total Revenue expend'd	1[illegible]	[illegible]	[illegible]	[illegible]	[illegible]	[illegible]
NET REVENUE ..	144,[illegible]	[illegible]	[illegible]	[illegible]	17[illegible]	[illegible]
Deduct Int'st on Bonds, Loans, &c. ..	71,7[illegible]	[illegible]	[illegible]	[illegible]	[illegible]	[illegible]
" Amalg'n company.	......	......	[illegible]	[illegible]	11,[illegible]	[illegible]
" Loss on Steamers	......	12,[illegible]	[illegible]	......	......	......
" Sinking Fund—Gov't Loan..	[illegible]	10,7[illegible]	[illegible]	......	......	......
" Renewal of Rails.	......	......	......	17,[illegible]	......	[illegible]
Deduct'ns from Net Rev	[illegible]	121,[illegible]	121,[illegible]	1[illegible]	79,737	[illegible]
Amt. available for Div'd	[illegible]	115,[illegible]	1[illegible]	[illegible]	[illegible]	[illegible]
Div'd per cent. per ann	6	5	[illegible]	[illegible]	[illegible]	[illegible]
Amt. carried forw'd	[illegible]	[illegible]	[illegible]	[illegible]	[illegible]	[illegible]
Miles open	[illegible]	[illegible]	[illegible]	[illegible]	[illegible]	[illegible]
Ost'y working expenses —per c'tage on earn'gs	[illegible]	[illegible]	[illegible]	[illegible]	[illegible]	[illegible]

GREAT WEST

ROLLING STOCK, &c.	1860.	1861.	1862.	1863.	1864.
Number of Locomotives	94	94	94	94	94
" Passenger Cars..	127	127	127	127	127
" Freight	1255	1255	1255	1255	1255
" Baggage and Mail	20	20	20	20	20
" Express.........	....	....	....	....	....
" Gravel	120	120	120	120	120
Mileage of Locomotives	1,650,345	1,829,964	1,964,372	1,907,993	1,900,190
" Passenger Cars...	3,446,744	3,538,075	3,863,894	4,091,745	4,384,097
" Freight Cars	10,909 966	13,446,186	16,197,862	14,409,472	12,364,018
Passengers carried .....(No.)	525,692	527,088	572,751	637,218	668,272
Freight, " ....(Tons)	374,088	441,515	540,632	523,221	476,498
REVENUE.	£ Stg.	£ Stg.	£ Stg.	£ Stg.	£ Stg.
Passengers....................	207,350 7 0	184,730 2 2	200,013 0 8	225,551 12 7	261,272 12 5
Freight and Live Stock......	220,147 9 9	273,870 6 11	337,705 11 6	352,101 0 4	346,954 19 7
Mails and Express..........	14,710 0 8	15,209 2 2	16,932 9 11	16,803 7 5	18,133 8 10
Sundries	4,232 5 4	1,458 7 3	1,283 11 7	1,245 18 3	1,307 9 2
Total Receipts..........	447,324 12 6	475,267 18 6	555,533 18 8	594,701 18 7	627,669 10 0
EXPENDITURE.					
Maintenance of Way, Works and Stations............	20,723 7 6	37,606 17 10	33,834 17 0	35,434 15 8	36,838 5 1
Locomotive Power..........	79,648 19 6	80,759 6 3	83,783 12 0	80,796 16 1	79,020 2 1
Repairs and Renewals of Cars	18,992 0 2	22,444 13 0	28,934 2 0	35,193 13 10	35,955 11 7
Transit Expenses............	40,861 15 0	39,673 10 0	40,809 16 11	40,412 9 8	45,947 19 8
Merchandize Expenses	37,476 15 6	43,432 8 8	43,228 3 9	48,088 19 3	49,788 1 0
General Charges	14,209 6 0	16,643 9 5	15,136 5 0	15,765 5 11	15,292 2 0
Total ordinary working Exp's	220,977 4 8	240,560 10 2	250,806 18 2	255,714 0 5	262,792 1 5
Taxes, Insurance, &c.......	5,443 11 1	5,233 15 3	6,037 8 0	7,188 9 1	5,596 13 6
Suspension Bridge Rent.....	9,246 11 6	9,246 11 6	9,246 11 6	9,246 11 6	9,246 11 6
Total Rev. Expenditure	235,667 7 3	255,041 2 11	265,250 17 8	272,140 1 0	277,625 16 5
NET REVENUE	211,657 5 3	220,226 16 7	290,282 16 0	322,552 17 7	350,043 13 7
Add Surplus from prev. year	23,813 3 7	15,790 16 0	14,054 6 6	3,348 12 4	4,129 9 8
" Div. on Detroit & Milw. Pref. Shares	....	....	....	....	....
" Profit on Galt & G. R'y	....	....	....	....	....
" Other Accounts........	....	....	....	....	....
	235 470 8 10	236,017 10 7	304,337 2 6	325,901 9 11	354,173 3 3
Deduct Interest on Bonds, Loans, &c., &c....	108,545 8 8	105,423 3 1	105,228 0 0	102,906 4 2	101,584 4 9
" Discount on Am. C'y.	....	292 16 2	47,831 10 1	114,085 5 1	152,016 16 5
" Loss on E. & N. R'y.	....	....	....	....	....
" Renewals of Rails ..	43,258 6 6	52,900 12 0	71,590 12 8	70,448 16 7	62,403 8 11
" Real. Fund Ferry strs	....	....	....	....	....
" Other Accounts	17,926 14 10	13,143 18 7	9,076 7 10	9,140 1 3	....
	169,730 10 0	171,820 9 10	233,726 19 7	296,580 7 1	316,004 5 1
Amount available for Div...	65,739 18 10	64,197 0 9	70 610 2 11	29,321 2 10	38,168 18 2
Dividend per cent. per annum	1.50	1.50	2	.75	1.00
Capital Expended...........	£4,726,048	£4,769,662	£4,781,728	£4,807,847	£4,842,140
Miles open	345	345	345	345	345
Ordinary Working Expenses, per cent. of Earnings...	50.51	51.44	....	43.01	41.86
Do. do., Cost per Train Mile	3s. 8d.	3s. 2d.	3s 0d.	3s. 7d.	3s. 0d.

ERN RAILWAY.

[illegible]	[illegible]	[illegible]	[illegible]	[illegible]	[illegible]
94	94	94	[illegible]	[illegible]	[illegible]
157	[illegible]	[illegible]	[illegible]	[illegible]	[illegible]
[illegible]	[illegible]	[illegible]	1871	[illegible]	[illegible]
[illegible]	[illegible]	[illegible]	[illegible]	[illegible]	[illegible]
....	....	15	15	[illegible]	....
[illegible]	[illegible]	[illegible]	[illegible]	[illegible]	[illegible]
1,[illegible],670	[illegible]	2,[illegible],794	2,[illegible],776	2,757,[illegible]	[illegible]
4,[illegible]	4,[illegible]	6,[illegible]	6,570,[illegible]	6,[illegible]	6,[illegible]
11,[illegible]	11,114,[illegible]	14,[illegible],617	17,[illegible]	20,[illegible]	[illegible]
714,[illegible]	[illegible]	718,771	704,[illegible]	797,[illegible]	[illegible]
[illegible]	[illegible]	[illegible],773	[illegible]	773,[illegible]	[illegible]
£ s. d.	£ s. d.	£ s. d.	£ s. d.	£ s. d.	£ s. d.
313,[illegible] 12 7	316,[illegible] 4 8	312,[illegible] 5 10	311,[illegible] 17 9	317,[illegible] 9 9	[illegible] 1
[illegible] 12 7	[illegible] 5 6	412,[illegible] 8 5	[illegible] 19 3	[illegible] 6 9	[illegible] 6 11
37,919 8 [illegible]	[illegible] 15 0	[illegible] 9 2	27,172 11 8	[illegible] 11 5	21,[illegible] 9 10
1,[illegible] 10 [illegible]	2,127 9 11	1,915 15 3	1,[illegible] 12 5	1,[illegible] 14 10	1,[illegible] 9 4
671,[illegible] 14 [illegible]	674,673 14 8	706,516 14 11	779,[illegible] 0 7	[illegible] 9 9	[illegible] 14 8
37,[illegible] 7	61,[illegible] 3 2	[illegible] 6 5	[illegible],244 5 8	166,[illegible] 3 5	170,[illegible] 7 9
73,476 3 [illegible]	70,[illegible] 14 7	[illegible],341 8 2	109,214 [illegible] 4	111,[illegible] 5 0	[illegible] 7 7
[illegible]	[illegible] 11 6	[illegible] 8 9	[illegible],477 6 10	[illegible] 7 3	[illegible] 6 0
[illegible] 1 [illegible]	[illegible] 1 8	[illegible],796 3 5	[illegible] 17 9	[illegible] 5 5	[illegible] 0 [illegible]
[illegible] 14 [illegible]	87,[illegible] 17 5	[illegible] 14 8	70,[illegible] 11 6	72,[illegible] 17 9	[illegible] 9
16,[illegible] 3 [illegible]	18,472 7 4	16,[illegible] 9 4	17,[illegible] 1 8	[illegible] 6 5	17,[illegible] 5 11
264,[illegible] 4 [illegible]	301,[illegible] 15 8	300,[illegible] 9 9	[illegible] 16 9	[illegible] 10 2	[illegible] 10 7
5,910 2 [illegible]	6,[illegible] 3 6	6,621 16 3	6,[illegible] 8 2	7,[illegible] 10 9	7,[illegible] 11 2
9,[illegible] 11 [illegible]	9,[illegible] 11 6	9,[illegible] 17 9	9,[illegible] 16 10	9,[illegible] 2 2	9,[illegible] 15 8
279,[illegible] 17 11	317,[illegible] 10 8	[illegible],485 3 9	[illegible],214 3 9	478,[illegible] 12 1	[illegible] 5 11
[illegible],179 14 [illegible]	357,[illegible] 4 0	390,[illegible] 16 2	361,746 16 10	[illegible] 16 11	[illegible] 10 3
4,870 18 7	11,[illegible] 6 0	1,128 13 9	2,[illegible] 6 5	2,[illegible] 17 1	2,[illegible] 5 8
....	....	....	5,811 2 2	16,874 15 8	....
....	....	674 6 1	....	[illegible] 9 5	1,[illegible] 14 1
11,377 8 [illegible]	....	[illegible] 16 4	....	....	4,[illegible] 3
[illegible],127 16 [illegible]	[illegible] 10 0	[illegible],048 11 4	369,[illegible] 5 5	334,[illegible] 16 10	[illegible] 19 9
100,[illegible] 13 1	98,330 1 4	101,[illegible] 19 5	104,[illegible] 2 8	91,[illegible] 14 1	[illegible] 9 0
[illegible] 4 [illegible]	[illegible] 18 11	119,[illegible] 6 9	124,110 6 2	[illegible] 19 6	[illegible] 16 0
....	613 6 0	2,467 4 10	1,[illegible] 10 2	2,[illegible] 2 10	[illegible] 4 0
[illegible] 9 [illegible]	[illegible] 10 6	....	....	....	....
....	....	....	5,000 0 0	3,[illegible] 0 0	[illegible] 0 3
....	6,647 0 1	6,219 3 6	6,[illegible] 17 4	1,[illegible] 0 0	[illegible] 2 7
[illegible] 6 [illegible]	719,[illegible] 16 10	[illegible] 14 6	[illegible],491 16 4	[illegible] 14 7	[illegible] 10
[illegible] 11 [illegible]	140,641 13 2	134,[illegible] 16 10	177,[illegible] 7 1	[illegible] 0 3	[illegible] 5 11
3.50	4.[illegible]	4.25	[illegible]	4.[illegible]	[illegible]
£4,[illegible]	£[illegible],691,[illegible]	£[illegible],111,[illegible]	£[illegible]	£[illegible]	£[illegible]
[illegible]	345	[illegible]	[illegible]	[illegible]	[illegible]*
[illegible]	44.73	47.[illegible]	51.[illegible]	[illegible]	[illegible]
[illegible]	[illegible]	[illegible]	[illegible]	[illegible]	[illegible]

* Glencoe and Buffalo extension, [illegible] miles under contract.

† Including renewals each subsequent year.

Total length of Main Line,—Niagara Falls to Windsor—229 miles.

Branches—Hamilton to Toronto	38	miles
Harrisburgh to Galt	12	do
Komoka to Sarnia	52	do
Wyoming to Petrolia	5	do
Total	106	miles.

Leased Branches—Galt and Guelph Railway (Galt to Guelph)	15½	do
Erie and Niagara (Fort Erie to Niagara)	30½	do
Wellington, Grey and Bruce (Guelph to Alma)	23½	do
Total of leased lines	69½	miles.

Sidings 79 miles.

Total length of railroad 484½ miles.

375 miles of the track are of iron and 40 miles of steel rails.

DIRECTORS.—Thos. Dakin, Thos. Faulconer, Paul Margetson, George Smith, London, Eng. ; John Fields, Manchester, Eng. ; Gilson Homan, Kirstall, Eng. ; Wm. Weir, Glasgow, Scotland ; Hon. Wm. McMaster, Toronto, Can. ; Hon. John Carling, London ; Donald McInnes, Hamilton, Can. ; M. K. Jessup, New York.

PRESIDENT,—Thos. Dakin, London, Eng.

SECRETARY—Brackstone Baker, London, Eng.

CHAIRMAN OF THE CANADIAN BOARD.—Hon. Wm. McMaster, Toronto.

OFFICERS.—W. K. Muir, General Superintendent, Jos. Price, Secretary and Treasurer, Hamilton ; Wm. Wallace, London, Ont., Assistant Superintendent ; Geo. Lowe Reid, Chief Engineer, Hamilton ; W. A. Robinson, Mechanical Superintendent, Hamilton ; J. Metcalf, Auditor, Hamilton ; A. E. Irving, Solicitor, Hamilton.

Principal Office, Hamilton, Can. London Office, No. 126 Gresham House, Old Broad St., E. C.

NOVA SCOTIA RAILWAY.

In the Province of Nova Scotia the construction of railways was first authorized by an Act of the Legislature, passed 31st March, 1854. During the same year another Act of that body authorised the issue of Provincial six per cent. debentures, having twenty years to run, in order to raise the necessary capital to proceed with the work of construction determined upon. These bonds were mostly sold in London, through Messrs. Baring Bros. & Co.; the Hon. Joseph Howe having been sent thither as a delegate with that object in view; a small amount found purchasers in the Province. It was provided that the proposed railways should be constructed under the supervision of one or more Commissioners, who were empowered to draw on the Receiver General for the monies disbursed to the contractors. They were restricted to the expenditure of $800,000 in any one year, beyond which amount they could not incur any liabilities.

The first sod of the Nova Scotia Railway—the first constructed in that Province—was turned at Richmond, on the 13th June, 1854. Sixty-one miles of railway to Truro were completed by the 15th of December, 1858, and the Windsor branch of the same road by June 3rd, 1858. An extension from Truro to Pictou on the Gulf of St. Lawrence, fifty-two miles in length, was afterwards built and opened for traffic on the 31st of May, 1867, making in all 145 miles of railway. The Windsor Branch, 32 miles, extends westward from Halifax to Windsor on the Bay of Minas, connecting with the Bay of Fundy. The total cost of the Railway, with equipment to 30th June, 1868, was $6,[illegible],647.69; and the total amount expended on construction account alone up to the 30th June, 1869, was $6,721,254.30.

The Pictou extension was surveyed by Mr. Sandford Fleming, C. E., and estimated to cost, including rolling stock, $2,314,500. Some of the original contractors abandoned their contracts, and work proceeding very slowly, the Government took the work out of their hands, and re-let the whole to Mr. Fleming for the sum of $2,116,500. The road was satisfactorily completed within the time specified, under the superintendence of another engineer. This extension cost to the 30th of June, 1868, the sum of $2,331,867.28.

The maximum grade on the whole line is 70½ feet per mile; minimum radius of curvature 792 feet. The iron rails laid down on forty-four miles of the main line are of the H pattern, or double headed, weighing 63 lbs. to the yard, supported at intervals of 2½ feet on cast iron chairs spiked into ties 10 feet long, ten inches wide, and 5 inches thick; the rails are secured to the chairs by wooden side keys. On another section of 17 miles, rails of the T pattern were laid 63 lbs. to the yard, and spiked into the sleeper. Between Truro and Pictou a T rail weighing 56 lbs. to the yard was laid, secured at the joint with a steel scabbard into which the rail fits, which proved to be very superior description of fastening. The H rail turned out very unsatisfactory. The gauge is 5 feet 6 inches. The bridges have iron girders resting on stone piers.

In 1868 the working expenses were three-fifths of one per cent. more than the gross receipts, against 87 per cent of the receipts in 1867. The earnings per mile in 1869 were $1,870 60, against $1,751 68 in the previous, year, and the operating expenses $1,852 14 and $1,762 18 for 1869 and 1868 respectively. The speed of the mail or passenger trains is about 10 miles an hour, and freight or mixed trains 13 miles.

In the official report of this railway for 1869, it is stated that "The trains with one exception, have been run with regularity; this delay occurred on the 8th of March, during a heavy snow storm, which continued for two days; the train from Pictou to Halifax was detained one day, and the train from Halifax to Pictou seven hours." Coal is being used for fuel, and is found much cheaper than wood. Of 24 engines purchased for the use of the railway, about one half were made in Glasgow, four in Kingston, Ontario, and the balance in the United States. The weight of the engines with tenders, light, ranges from 74,000 to 94,000 lbs.

OFFICERS.—George Taylor, General Superintendent; A. W. Hale, Engineer; Thomas Foot, Accountant; Wm. Johnston, Locomotive Superintendent; William Marshall, Road Inspector.

CHIEF OFFICE AND ADDRESS.—Halifax, N. S.

NOVA SCOTIA RAILWAY.

FISCAL YEAR ENDS 30th JUNE.

Rolling Stock, &c.	1860.	1861.	1862.	1863.	1864.	1865.	1866.	1867.	1868.	1869.	1870.
No. Locomotives	20	20	20	20	20	20	20	22	19	20	21
" Pass'gr Cars	18	18	18	18	18	19	21	26	25	23	26
" Freight	147	142	144	146	148	162	212	256	232	236	228
" Baggage &c.	....	....	....	....	....	....	....	5	2	3	6
" Coal Cars...	....	....	....	....	....	....	....	....	61	61	101
" Gravel	15	13	13	13	13	13	12	12	11	10	10
Mile'ge locomotvs	135,854	155,790	132,300	....	171,181	179,261	185,753	199,005	250,021	328,404	[illegible]
No. Pass'grs car'd.	88,044	81,359	96,020	101,188	....	139,890	149,333	106,879	160,871	187,769	236,402
Freight, Tons...	....	....	42,135	56,471	....	62,503	70,257	58,415	91,770	108,330	136,506
	$	$	$	$	$	$	$	$	$	$	$
Passenger Traffic.	61,705 38	56,854 56	63,489 84	66,580 70	56,878 47	82,073 21	86,711 84	96,006 99	114,467 44	127,132 94	130,968 00
Freight.......	50,085 54	59,143 63	69,209 40	77,800 64	63,724 35	98,666 24	107,929 42	96,006 61	136,944 50	133,897 12	146,581 68
Express ......, Mails ...., Sundries	4,271 97	4,829 47	6,407 33	5,023 08	2,151 73	3,214 37	4,004 93	2,492 74	2,682 22	11,908 28	9,186 06
Gross Earnings ..	116,742 09	120,917 66	139,106 71	149,674 42	121,754 45	183,953 82	199,730 19	185,098 34	253,994 16	272,237 41	275,867 13
Gross expenditure	96,472 26	94,114 88	101,936 23	137,962 58	98,242 90	150,068 88	165,571 8	136,388 98	256,330 51	268,509 57	301,038 30

THE NORTHERN RAILWAY:

The Toronto, Sarnia and Lake Huron Railway Company was chartered, (12 Vic. c. 196), August 29, 1849, with a capital of £500,000, in £5 shares. The road was to run from the City of Toronto to some point on the southerly shore of Lake Huron, touching at the town of Barrie on the way. The survey of the road was to be deposited within three years and the road to be completed within ten years. The Company was authorized to raise the amount of the stock either by subscription or by lottery, but the whole amount of the proceeds were to be devoted to the purposes of the railroad. The lottery scheme was never put into practice. By 13 and 14 Vic. c. 131, the title was changed to the Ontario, Simcoe and Huron Railroad Company; and in 1858, the name was again changed to "the Northern Railway of Canada." Toronto was, after the passing of the 13 and 14 Vic. c. 131, no longer necessarily the starting point. Authority was given to commence the road at any point on Lake Ontario, west of the township of Darlington.

By 16 Vic. c. 244–5, authority was given (1853) to the Company to construct a branch line to the eastern shore of Lake Huron, not further south than the southerly limit of the township of Saugeen, and to construct at such point a harbour and other necessary werks; to increase the stock to £750,000 and to borrow a further sum of £300,000. There was no clause guarding, in express terms, against this new loan taking precedence of the Province claim; but there was a singular clause against anything in this or any other Act being construed as binding the Province to guarantee the interest of any loan to be raised or debenture to be issued by the Company.

The length of the road is ninety four miles, besides sidings which extend to something like fifteen miles. There were besides a few miles of double track. The minimum radius of curvature is 1,432 feet, and the maximum grade going north is 60 feet; going south 52 feet 8 inches.

The first section of the road, from Toronto to Aurora, 30 miles, was opened to the public on the 16th May, 1853; the next section to Bradford, on the 13th June, 1853; the third section to Barrie, on the 11th October, 1853; the branch to Bell Ewart, a mile and a half, on the 3rd May, 1854; and before the end of that year, the whole line was open for traffic. The first sections were opened before the ballasting was done; and the work was afterwards performed when the road was in operation.

With a view of controlling the navigation of Lake Simcoe, the Directors purchased the steamer Morning and the wharves at Orillia and Bradford, and afterwards built the steamer J. C. Morrison.

The original contract with Storey & Co., for construction, was for £579,-

176 8s. 6d., and a supplementary contract for locomotive stock, general rolling stock, way station service, terminal depot service, harbour service, and steamboat service brought the amount up to £70[illegible] 1s. 2d. currency.

Mr. Brunel, C. E. estimated the revenue at £1[illegible] currency per annum, giving a net revenue of £[illegible], equal to eight per cent. on the entire cost of the road, wharves and harbours.

The Company received from the Government, in the shape of guarantee, £475,000 sterling; and it paid the interest on the Government bonds issued on its behalf, up to the 1st January, 1858—the original capital account being open. The total amount paid under this head, with commission, is £47,-[illegible] 18s. At first the Province had a first lien on the whole of the Company's line of railway from the City of Toronto to Collingwood harbour on Lake Huron, and all the ground belonging to the said Company, enclosed or to be enclosed, and lying between the said termini, together with all the station-houses, wharves, store houses, engine houses and other buildings thereon erected." Default in the payment of interest on the Government bonds was first made in the amount that became due after the 1st January, 1858, and nothing further was ever paid. In other words, so long as there was original capital out of which to pay the unearned interest it was paid, but never afterwards.

The want of connection with the Northern terminus, at Collingwood, was early felt, and in 1855, the Company, with a view of developing the business of the line, entered into contracts for a tri-weekly line of steamers between that port and Lake Michigan ports, and a weekly line to Green Bay. Five first class steamers were employed and the charter money paid to them was £21,750 currency. In 1855, the income of the Company was £9[illegible],872 12s. 8d currency, and the expenditure £130,698 2s. 1d., showing a loss of £[illegible]-9s. 6d. Next year there was paid on account of the steamboat contract only £6,750. On the 24th September, 1856, the steamer Niagara, one of the line, was lost near Port Washington, with many lives and a cargo of freight. In 1859 this line of steamers had become self-sustaining, and the Company derived a profit of over $10,000 from the connection. They then resolved to entertain no proposition for future connections with the Upper Lakes, which would involve any subsidy or guarantee. This determination, together with the heavy work of renewal on the line, led to a suspension of the steamboat organization between Collingwood and Chicago, causing a trifling falling off in the through trade in 1861, but it was scarcely appreciable, being less than $2,000, so nearly did volunteer competition, both of sail and steam vessels fill up the void. During the summer of 1862, but not till the first month of the navigation was over, four first class propellers maintained and strengthened the reputation of the route. This season vessels were scarce on Ontario, and the Company suffered seriously with its connections at that end, delays and accumulation of freight rendering it liable for damages. These difficulties were finally overcome by securing the services of two propellers for the remainder of the season, on favorable terms. In 1863 the Company found the American carrying trade too fluctuating and hazardous to justify its making any special arrangement with regard to it, and from that time this policy has been carried out, in connection with the special development of the local traffic.

The Company owned steamers on Lake Simcoe, which it chartered to other parties in the spring of 1866, but the arrangement fell through by August, and the Company ran them for the remainder of the season.

In 1855–56 the expenditure was £5,475 over earnings. The passenger trains ran at the rate of 25 miles an hour, when in motion, and 20 miles including stoppages, and the express trains ran five miles an hour faster; freight trains 15 miles when in motion and 12 miles including stoppages.

In 1857, "An Act to amend the charter of the Ontario, Simcoe and Huron Railroad Union Company," (20 Vic. c. 143), enacted that so long as the City of Toronto shall hold stock to the amount of £25,000, it may appoint one of the Aldermen a director of the Company, and the County of Simcoe may, on the same condition, also nominate a representative at the Board.

In 1858, (Vic. 22 c. 117) the name of the Company as already stated was changed to "The Northern Railway Company of Canada," authority was given to call in all the outstanding bonds, exclusive of those granted to the Government, and to issue to the holders other bonds, in lieu of them; and to issue £200,000 six per cent. sterling bonds for the purpose of funding the floating debt, to extend the works and put the road into efficient working order.

At this period, the order of priority in the capital account of the Company was: Government lien £475,000, with (August 1 1859) £116,375 arrears of interest thereon, making a total under this head of £591,375 stg. Next came Company's bonds £243,739 14s. 6d., with unpaid interest theron, £43,434-8s., a total of £287,174 2s. 10d. Third amount required to cover floating debt and place the road in an efficient condition, £250,000. And there had been paid on stock subscriptions £169,276 8s. 3d., making a total capital of £1,297,825 11s. 1d.

In 1859, an Act was passed, vesting in the crown all the real and personal property of the Company, for certain purposes therein set forth. The Government was to raise an additional sum of $60,000, which, added to the then existing claim of the Province, was to be a first charge on the proceeds of the sale of the road, the day of this sale being fixed on the then ensuing first of August, and the proceeds of the sale were to be distributed among the creditors, including the Province, in the order of their priority, claims of equal rank being paid *pro rata*, if there should not be sufficient to pay them all. The Government took authority to cause the railway to be worked through the intervention of the Company or any other party or parties; but the surplus after paying working expenses and interest on the claim of the Province, was to be paid over to the Company. The Government might become a purchaser at the sale, for an amount not exceeding its claim. The Government might treat with the Company or its bondholders, or both, for the transfer to them of the railway and its appurtenances, and allow the new proprietors to issue £250,000 of preferential bonds for the repair of the road and the payment of the Company's debts, the $60,000 to be advanced, meanwhile, by the Government to be repaid before any other outlay was made.

In pursuance of the large additional powers given to the Government, an order in Council was passed in May, 1859, in which the Minister of Finance declared there was no reasonable hope that any parties would be found to offer any considerable sum of money for the railway, if sold, in which case the Province would either be required entirely to sacrifice the whole of their claim or to assume the works themselves, and to advance from Provincial funds the sums required to maintain the line." He took the ground that in any case, it was not desirable to increase the debt of the Province for the purpose of aiding the road; that, for many reasons, it was not desirable, except as a last resort, to make use of the power of absolute sale. He therefore recommended that the whole property be revested in the Company, on the following conditions: The Company should be allowed to issue £250,000 stg. of first preference bonds, which would displace the Provincial lien; that the advance made by the Government in 1856 of £10,000 stg., with interest thereon and such amount as the Government might now spend in repairs, be repaid before the 25th December 1859; that the Company repair the line and complete the rolling stock; that £50,000 of second preference bonds be handed to the Government in payment of past interest on the Provincial claim. In consideration of the fulfilment of these conditions, the Government was to grant priority of dividends over the Provincial claim to the amount of the existing debenture debt of the Company, $1,185,884. A curious provision was inserted with regard to the mortgage bonds, amounting to $192,[illegible], as some of them had been sold below par, the priority, in their case, was only to extend to the amount the Company had received, a principle which would certainly be pronounced false if applied to the National debt. But the claim of these bondholders as against the Company was not to be diminished. The interest due on the existing bonded debt was to be postponed to and rank after the Provincial claim of £375,000 stg. Second preference six per cent. bonds were to be issued to pay off, by way of exchange, the existing bonded debt and for £[illegible],000 back interest owing to the Province, and to have priority over the Government lien. The order of priority, in which the future earnings of the railway were to be disposed of, subject to the above conditions, was: 1st, Working expenses, repairs and management. 2nd, Interest on first preference bonds. 3rd. Interest on second Preference Bonds. 4th, Interest on the Provincial Lien of £475,000 stg. 5th, Interest on arrears of interest due the Province. 6th. Interest on the proportion of the mortgage bonds entitled to this priority and on the arrears of interest due on the pre-existing bonded debt up to the date of the issue of the second mortgage bonds. 7th. Interest on the share capital.

The priority of capital now stood: 1. First Preference Bonds, £[illegible],000. 2. Second Preference Bonds, £[illegible],189 14s. 6d. 3. Government lien, £475,000. 4. Balance of interest arrears due the Province, £[illegible]. 5. Interest arrears on Company's bonds, £43,481 8s. 4d. 6. Stock subscriptions amount paid, £129,276 8s. 3d. Total £1,[illegible],925 11s. 1d. sterling.

The "Northern Railway Act of 18[illegible]" empowered the Company to issue third Preference Bonds (class A.) to the amount of £[illegible],000 stg., and to "expend the proceeds thereof in the construction of elevators, the increase and extension of rolling stock and other equipment works for the accommodation and facilities of the traffic." The new elevator constructed at Toronto has a storage capacity of 275,000 bushels, and can elevate and ship 20,000 bushels an hour. The elevator wharf, sunk in 15 feet of water, is 400 feet long and 70 wide, and can store three million feet of lumber for shipment. A new elevator at Collingwood, nearly as large as this, was included in the works constructed by these bonds :it will be completed by the 10th August. When the road was first built, a breakwater and wharf were constructed at that port, for the safety and convenience of the traffic connections. The elevator previously used by the Company at Toronto was burned down in the early part of 1870. A similar casualty happened some years before, in the burning of the Company's steamer, "J. C. Morrison," on Lake Simcoe. The policy of owning steamers on that lake has long been discontinued.

This railway has been of immense benefit to Toronto and the whole northern country. It has hitherto been the only road terminating at Toronto, and the facilities it has afforded have opened up a new and large lumber trade, on the Georgian Bay.

When Mr. Cumberland became Managing Director in 1859, he changed the whole policy on which the road had been worked. Large gross receipts, if they left no profit, had no charm in his eyes. He found the through traffic had been carried at a loss ; at a loss so great that in the previous year, it had more than eaten up all the profits of the local traffic. He informed the proprietors of his intention, and warned them not to be alarmed if they found a considerable decrease in the gross revenue. He intended to do none but paying business ; to touch nothing that did not leave a profit. How this policy succeeded the following table will show. In 1858, there had been a positive loss on the whole business ; in 1859, under the new policy, the total receipts showed a decline of nearly twenty thousand dollars ; but this diminished revenue brought with it a profit of nearly forty-three thousand dollars. The working expenses still bore a very large proportion to the revenue, over 82 per cent. This item has undergone a constant reduction, till it is now only a fraction over 58 per cent. Every possible encouragement is given to the development of local traffic ; sidings being put in wherever there is a promise of business to warrant it. This policy, which has been eminently successful, might be impossible in a line of great length, where competition rates are fixed by the cost of carrying on the most favorable route ; but for the Northern there cannot be a question, it has proved the true policy, as tested by the touchstone of success.

DIRECTORS.—Hon. John Beverley Robinson, President, Toronto ; John A. Chowne, Esq., Vice-President and Chairman of London Board, London, England ; Fred. W. Cumberland, Esq., Managing Director, Toronto ; Angus Morrison, Esq., M. P., Toronto ; Wm. Elliot, Esq., Toronto ; Henry Wheeler Esq., London, Eng. ; H. M. Jackson, Esq., London, Eng. ; W. D. Ardagh,

Esq., Barrie, Warden County Simcoe; Ald. J. J. Vance, for Corporation of Toronto.

Officers.—Fred. W. Cumberland, General Manager; Thomas Hamilton, Secretary and Accountant; C. W. Moberly, Chief Engineer; Francis Tustin, Mechanical Superintendent; John Harvie, Train and Traffic Master; Clarke, Gamble, Q. C., Geo. D'Arcy Boulton, Solicitors; Wm. Gamble, Ed. B. Osler, Auditors.

Chief Office for Canada.—Toronto, Ont.

London Agency.—Messrs. Cutbill, Son & De Lungo, No. 166 Cannon Street, London, E. C.

NORTHERN RAILWAY.

Table exhibiting the Earnings, Expenses, Profit, and Per Centage of Working Expenses to Gross Receipts, until the date of re-organization.

	1854.	1855.	[illegible]	*[illegible]	†1857	[illegible]	[illegible]
Earnings	114,887 [illegible]	2?3,4?? [illegible]	2?0,??? [illegible]	[illegible]	31?,?91 [illegible]	[illegible]	[illegible]
Working Exp.	85,0?? [illegible]	167,377 [illegible]	2?6,0?5 [illegible]	146,220 91	2?0,??? [illegible]	351,717 [illegible]	1?7,??? [illegible]
Profit	29,7?0 7?	46,0?? [illegible]	....	4?,0?? [illegible]	[illegible]	....	[illegible]
Loss	....	....	5,478 1?	....	....	15 [illegible]	....
Per centage of working exp. on gr. receipts	74.67	78.51	101.80	81.??	79.70	100.0?	82.?4

* Half-year ending December. † Year changed from 30th June to 31st Dec.

NORTHERN

Exhibit of the Rolling Stock, Mileage, &c., also Revenue, Expenditure and

ROLLING STOCK, &c.	1859. 5 Months.	1860.	1861.	1862.	1863.
Number of Locomotives	17	17	17	18	18
" Passenger Cars, 1st class..........	13	13	13	18	19
" Freight (Box)....	116	116	108	117	117
" Baggage, Express and Mail......	4	4	4	6	4
" Platform	160	160	175	178	186
" Other............	15	15	15	16	14
Mileage of Locomotives......	......	300,280	353,122	347,249	359,674
Number of feet sawn lumber, carried B. M...	......	......	8,657,303	27,626,000	25,916,000
" Cubic feet square timber........	......	......	34,651,692	17,938,000	21,164,180
" Passengers carried	......	91,582	100,618	101,629	107,882
Freight, Tons	......	125,345	145,754	174,345	145,994
REVENUE.	$	$	$	$	$
Passenger earnings..........	33,014 37	68,741 49	94,072 96	96,678 14	102,147 66
Freight, Local	50,763 48	196,085 80	260,434 98	209,065 69	275,462 09
" Through..........	21,906 86	50,367 62	48,432 41	92,092 70	18,295 74
Mails	1,175 00	3,564 14	3,454 22	3,453 34	3,468 44
Express	......	......	......	....	......
Wharfage	1,074 53	1,424 60	928 76	1,028 93	1,808 86
Storage	571 32	2,210 42	2,836 58	2,112 28	4,104 16
Rents	......	......	.....	......	......
Other Sources	421 35	573 04	780 00	1,211 84	1,834 50
Total Earnings..........	108,926 41	332,907 01	410,939 01	406,238 02	408,066 55
EXPENDITURE.					
Permanent Way and Works..	22,615 18	65,494 22	74,001 76	55,125 30	46,866 91
Machinery and Rolling Stock	14,663 15	37,096 37	30,561 60	46,205 52	51,006 16
Operating	42,570 42	134,893 35	153,740 85	199,215 84	87,715 92
Other Expenses	18,697 39	22,962 62	11,655 36	8,106 34	32,950 08
New Works, Rolling Stock, &c.	......	......	. ....	......	8,740 12
Total Expenditure	93,596 14	260,406 46	278,968 80	308,653 08	226,878 74
NET REVENUE........	15,330 37	72,500 46	131,971 11	97,584 90	179,727 81
Add Unexpended Balance....	......	15,330 27	24,417 12	27,517 96	23,327 98
" Interest on Deposits, &c.	......	......	1,387 38	4,159 52	*12,873 64
Amount available for Int. Div.	15,330 37	87,830 72	157,775 61	129,262 44	215,929 43
D'ct 1st Preference Bonds....	......	25,301 80	43,048 60	64,527 18	73,000 00
" 2nd " "	....	38,111 80	80,369 63	40,807 00	82,825 80
" 3rd " "	......	....	......	......	......
" Appr't'n to special works	......	....	.....	......	19,340 86
" Discounts, Stamps, Arrears of Interest, &c..........	.. ...	....	1,659 12	600 33	4,396 61
Balance on 31st Dec. to next year....................	......	24,417 12	27,517 96	23,327 96	36,367 77
		87,830 72	157,775 61	105,934 46	215,929 43
Extension of works..........	*	....	......	8,106 34	8,340 1
Per centage of Working Exp. on gross receipts........	......	78.22	67.88	73 98	53.74

* To which is added $9,442 re-payment of accrued interest on bonds issued under Restoration contract.

RAILWAY.

Net Revenue and appropriations of the same, since the date of re-organization.

[illegible]	1855.	[illegible]	1857.	[illegible]	[illegible]	[illegible]
16	16	16	16	20	21	[illegible]
19	19	19	[illegible]	[illegible]	17	[illegible]
117	[illegible]	114	141	147	147	[illegible]
5	5	5	6	6	6	7
[illegible]	214	[illegible]	[illegible]	370	[illegible]	[illegible]
6	7	8	7	7	14	[illegible]
[illegible],070	[illegible]	[illegible],913	464,[illegible]	[illegible]	862,731	[illegible]
[illegible]	97,[illegible],800	62,[illegible]	64,790,000	84,964,000	[illegible],000	117,[illegible]
3,[illegible],673	2,[illegible]	2,147,[illegible]	2,718,[illegible]	1,616,[illegible]	[illegible]	1,672,[illegible]
[illegible]	[illegible],872	187,[illegible]	1[illegible],140	[illegible]	14[illegible]	[illegible]
[illegible]	119,671	174,[illegible]	390,[illegible]	794,[illegible]	[illegible]	[illegible]
9	9	8	9	9	9	9
164,772 [illegible]	16[illegible]	123,[illegible] 08	18[illegible] 92	146,[illegible] 82	160,[illegible] 20	172,[illegible] 96
[illegible]	[illegible],912 [illegible]	300,113 [illegible]	[illegible]	304,206 36	679,[illegible] 12	[illegible] 10
14,[illegible]	[illegible]	[illegible] 96	31,[illegible] 73	12,030 79	[illegible] 01	[illegible] 67
6,[illegible]	6,717 [illegible]	9,[illegible] 35	9,[illegible]	9,707 21	[illegible]	[illegible]
......	†12,[illegible]	......	......	......	[illegible]	2,[illegible] 75
[illegible]	4,011 [illegible]	[illegible]	4,[illegible] 81	6,[illegible] 04	6,811 [illegible]	[illegible]
[illegible] 63	6,[illegible] 64	[illegible],614 02	[illegible],188 01	7,679 94	9,446 [illegible]	[illegible] 79
......	......	......	......	......	1,[illegible]	2,277 [illegible]
[illegible] 94	3,[illegible] 31	2,[illegible]	2,[illegible] 70	1,[illegible] 35	1,[illegible]	1,[illegible]
[illegible] 13	[illegible],745 [illegible]	512,874 08	561,370 26	[illegible],070 34	671,076 31	72[illegible] 88
99,[illegible] 18	84,[illegible] 30	104,614 [illegible]	[illegible] 34	102,363 97	111,[illegible]	[illegible]
[illegible]	61,516 [illegible]	[illegible]	76,517 06	67,[illegible] 97	84,[illegible]	[illegible]
[illegible]	[illegible]	[illegible],110 [illegible]	118,[illegible] 01	116,[illegible] 17	1[illegible] 36	[illegible]
[illegible] 74	[illegible]	[illegible] 00	46,[illegible]	[illegible],151 70	64,[illegible]	64,71[illegible]
81,[illegible] 07	51,619 [illegible]	27,[illegible] 56	[illegible] 85	[illegible]	161,122 [illegible]	[illegible]
[illegible],122 [illegible]	307,[illegible]	[illegible]	[illegible],442 75	[illegible] 29	[illegible],168 [illegible]	[illegible]
[illegible] 71	179,[illegible]	176,[illegible]	168,[illegible]	165,210 [illegible]	16[illegible] 11	171,[illegible]
[illegible] 77	[illegible]	[illegible] 77	46,[illegible]	[illegible],263 [illegible]	[illegible]	3,717 [illegible]
3,[illegible]	2,180 7[illegible]	2,130 74	3,130 74	3,[illegible]	2,016 [illegible]	6,[illegible]
[illegible]	216,[illegible] 37	216,[illegible] 75	164,164 60	169,[illegible]	[illegible],194 [illegible]	101,[illegible]
[illegible],000 00	73,000 00	[illegible],000 00	[illegible],000 00	73,000 00	[illegible],000 00	73,[illegible] 00
1[illegible]	[illegible]	[illegible]	[illegible]	82,[illegible] 00	124,[illegible]	[illegible]
......	......	......	......	......	4,[illegible]	14,[illegible]
......	......	......	......	......	.... .	
......	20,794 70	20,794 70	......	13 08	......	
[illegible]	[illegible] 77	46,[illegible] 26	[illegible]	[illegible]	8 917 [illegible]	[illegible]
[illegible]	216,[illegible] 37	216,770 75	[illegible],164 60	16[illegible]	[illegible]	101,[illegible] 34
[illegible] 07	51,470 [illegible]	27,[illegible]	[illegible]	[illegible]	16[illegible]	[illegible]
[illegible]	56.[illegible]	[illegible]	[illegible]	61.[illegible]	[illegible].57	[illegible]

†Arrears for postal service since re-organization.

LONDON AND PORT STANLEY RAILWAY.

This railway connects the City of London, Ontario, with Lake Erie, is 24½ miles long, with 3¼ miles of siding, and cost \$1,027,928.24. It was commenced in 1854, and completed in Oct. 1856; weight of rail per yard 56 lbs. Termini London and Port Stanley. Iron rails, wooden bridges and buildings.

It is owned principally by the City of London, the Counties of Middlesex and Elgin, and the town of St. Thomas. The amount of private stock held is \$27,750. The capital stock subscribed by the municipalities was as follows:—City of London \$220,000; County of Middlesex \$80,000; County of Elgin \$80,000; Town of St. Thomas \$8,500—Total \$388,500. The railway is now indebted to London as follows:—1st Mortgage Bonds \$175,400; Stock in Road \$220,000; Loans on 1st and 2nd Mortgage Bonds \$220,000; Interest &c., \$502,126.—Total \$1,342,248.

The building of this road was commenced with a view to the general advantage and improvement of the country interested rather than from any expectation of profits to be derived directly from its revenue. The amount of traffic which was anticipated has not been realized by its projectors; but it has called into existence a competition from the trunk lines which has caused a reduction of the freights on farm produce and merchandize of fully 50 per cent., thus answering the purpose for which the road was constructed, and indirectly has paid a large dividend on the amount invested.

From the geographical and local position of the line, its direct communication with the leading thoroughfares in the United States and Canada, forming as it does, junctions with the Great Western and Grand Trunk Railways at London, with the Canada Air Line and the Canada Southern at St. Thomas, and its terminus at Port Stanley being only about 80 miles from Erie and Cleveland and being situated immediately opposite the great coal fields of Pennsylvania and Ohio will form a main artery through which the coal traffic may flow into the Western part of the Province of Ontario. As the land becomes cleared and the sources from which fuel is at present drawn become impoverished the value of this Road is expected to increase in proportion. From the natural increase of traffic that is to be expected, the property must increase in value. Should the London, Huron and Bruce Railway for which a charter was granted last session, be carried out as there is every likelihood of its being, it also will form a valuable feeder to this road.

The cost of construction was from the nature of the country through which it passed necessarily very heavy, there being heavy cuttings, long embankmenced, and expensive bridges causing an increase in the cost of the road over the estimate of fully $400,000. The enterprise was therefore placed under great financial difficulties when opened, and having no source from whence relief could be drawn, it has had to struggle on amidst many difficulties ever since. By the economy of its management and the energy of its officers it has gradually surmounted its difficulties, and promises during the coming season to see its way clear to a brighter future. The road has been remarkably free from accident both to person and property. All its original stock of cars are still in good working order. During the 14 years it has been running only one train has been cancelled and all the trains have been run with uniform regularity.

This Road cultivates a large excursion business, forming as it does an outlet for the populations of the City of London and Town of St. Thomas, to visit and ruralize on the shores of Lake Erie, where the Company have extensive pleasure grounds. To encourage this class of traffic the rates are necessarily very low, not realizing half a cent per mile for each passenger; but the managers have found from a judicious attention to this branch of business that a considerable income may be realized with very little extra expense. Occurring at a season of the year when the freight traffic is light, without any extra staff or labor these excursions become at once profitable to the road and a boon to the public.

The road is 5 feet 6 inch gauge, but there is no doubt but it will soon have to be altered to 4 feet 8½ inches, as all its connections except the Grand Trunk are adopting that gauge.

From the last year's report we gather the following:

Earnings Passenger Traffic	$18,429.00
" Freight Traffic	22,326.73
" Mails and Express	1,769.11
Sundries	257.60
Total Revenue	$42,782.44
Total Working Expenses	30,024.00
Net Profit	$12,758.44

The train mileage was as follows:

Traffic	44,906 miles.
Wood, Gravel and Repairs	3,512 do.
Total train mileage	48,418 miles.

The cost for operating the road including repairs, renewals, Fuel, Traffic, Superintendence and all expenses 66½c. per train per mile.

GENERAL BALANCES.—Capital Stock $442,349.00; Bonds $490,696.00; Other accounts, &c., &c., $150,500.00.—Total $1,083,545.00

PER CONTRA.—Construction to 30th Nov. 1870 $947,340.00; Rolling Stock $90,579.00; Stock and Bonds $25,328.00; Current Assets $19,294.00.—Total $1,083,545.00.

LONDON AND PORT STANLEY RAILWAY.

Rolling Stock, &c.	1860.	1861	1862.	1863.	1864.	1865.	1866.	1867.	1868.	1869.	1870.
Number of Locomotives..........	2	2	2	2	2	2	2	2	2	2	2
" Passenger Cars.......	3	3	3	3	3	3	3	3	6	0	6
" Freight "	42	42	42	42	42	42	42	42	42	42	44
" Baggage "	2	2	2	2	2	2	2	2	2	2	2
Passengers carried (number)......	21,919	23,375	30,636	31,879	35,115	37,098	45,633	43,923	42,701	39,418	44,427
Freight, " (tons)............	16,730	20,450	21,750	20,838	23,291	18,222	22,819	25,493	23,868	22,328	23,831
Revenue.											
	$	$	$	$	$	$	$	$	$	$	$
Passenger Traffic..................	9,793	11,506	13,647	13,624	14,553	14,855	19,560	17,308	16,969	15,971	18,439
Freight Traffic....................	17,100	20,098	19,632	19,645	19,023	16,327	21,188	23,064	22,401	20,369	22,536
Express, Mails and Sundries....	2,277	1,982	1,692	2,031	2,956	2,008	2,246	2,314	2,334	6,088	8,064
Gross earnings................	29,170	33,586	34,971	35,300	36,532	33,190	42,994	42,686	41,704	42,428	49,039
Expenditure.											
Operating and other expenses....	21,713	24,160	35,670	26,049	25,726	26,043	31,507	30,834	36,463	29,617	30,293

Directors—(elected January 1871,) Murray Anderson, J. M. Cousins, James Egan, John B. Smyth, Thomas Arkell, Elisha Paul, T. M. Nairn, Robert Thompson, John Waterworth.

President.—Murray Anderson.

Superintendent, Treasurer and Engineer.—William Bowman, Esq

Chief Office.—London, Ont.

COBOURG AND PETERBOROUGH RAILWAY.

This road from its very first inception has been a constant series of mishaps, disasters and changes. It was constructed under a charter obtained in the year 1852, authorizing the building of the same from Cobourg to Peterboro'. The first sod was turned on the 9th February 1853 with a great parade, the citizens of Cobourg turning out *en masse*, and having a ball and torchlight procession in honor of the occasion. The contractor for the building and equipping of it was the late Samuel Zimmerman, at a nominal price per mile, but which amount was entirely lost sight of long before the road was half finished. The Contractor made demands which the Directors considered exorbitant and unreasonable, and to which they refused to accede. The following are extracts taken from the annual Report of the Directors to the Shareholders in February 1856. The report of that year commences thus: "The Directors appear before you to-day with their annual Report anxious and disappointed. The railway under their management contracted to be completed in July remains still unfinished." Again "if the Contractor's claims are paid in full according to the accounts put in, the cost of the road will greatly exceed the first estimates." It appears further by the same report that the claims put in by the Contractor were obliged finally to be paid, and the road was eventually taken off his hands in a half finished state, he having obtained nearly all the ready money the Directors had been able to obtain from the Municipal Loan Fund, besides bonds to the amount of sixty thousand pounds sterling.

The gauge of the road is 5 feet 6 inches, and it was equipped by the Contractor with three locomotives, two passenger cars, ten box cars and thirty platform cars. The Directors having got possession of the road then went to work to finish it, but were met at all points with almost insurmountable difficulties from the very imperfect manner in which the road bed and bridge across Rice Lake was constructed. A bridge of three miles in length across Rice

Lake built on piles not sufficiently driven or even properly stayed, half way between the towns of Cobourg and Peterboro', was one of the difficulties in the way, but nevertheless the road was so far completed as to be opened for traffic in the month of December 1854. The occasion was celebrated with much rejoicing by an excursion trip to Peterboro' ; but short was the gratification of the Directors, for the first winter's frost stopt all running of trains.

The expansion and contraction of the ice and consequent shoving was so great that it entirely destroyed the bridge, thereby stopping all running of trains for some considerable time. Indeed it was not till the following spring that the road was sufficiently put in a state of repair to recommence its business traffic. The road only 30 miles in length had by this time cost a sum of money falling not far short of $1,000,000, namely £125,000 currency, borrowed from the Municipal Loan Fund, and £100,000 sterling of bonds issued, besides private stock to the amount of about £4,000. The road was then run by the Board of Directors until the year 1857, the whole line not realizing sufficient to pay working expenses, and the interest on the sterling bonds, in consequence of the constant repairs required on the bridge.

In 1857, a proposition having been made by Mr. D. E. Boulton to lease the road, on terms which the Directors considered would enable them to meet their interest liabilities his offer was accepted, and he commenced working the road with a fair prospect of success, but just at that time Mr. Fowler had projected a scheme of building a branch line from the Port Hope and Lindsay Railway at Millbrook to the town of Peterboro, and which by Government aid he was enabled to complete. This so materially reduced the traffic of the Cobourg Road that Mr. Boulton found it impossible to meet the engagement he had undertaken. In August 1858, the Bondholders obtained an Act authorizing them to take possession of the Road, which they did, and put it under the management of Mr. J. H. Dumble who worked it until January 1860, when again a change took place. The Road then passed into the hands of Messrs. Covert and Fowler as lessees. These gentlemen contracted to make the bridge across the Lake a permanent structure by filling in an earth embankment, but when it was only partially finished, another change took place ; the work ceased, and the bridge, which was before in a dilapidated state, was left to its fate. The result was that the following winter in consequence of the movements of the ice and the spring floods the bridge took its departure and sailed down the Lake ; and thus terminated the existence of the Cobourg and Peterboro' Railway, so far as all connection between the towns of Cobourg and Peterboro' is concerned.

Application was again made to Parliament by the Bondholders for relief, in 1862, when an Act of amendment of the Charter was obtained. In the year 1865 it was again amended.

The Railway was finally sold to a Company for the lump sum of $100,000. Out of this unpaid liabilities for rights of way and certain privileged claims were paid off, an arrangement was made with the Bondholders for their payment in certain proportions, and all other and further claims and liabilities were wiped out. In 1869 an act was passed by the Ontario Legislature authorizing the amalgamation of the Cobourg and Peterborough Railway Company and the Marmora Iron Company. The same act authorized the amalgamated Company to issue debentures to the amount of $[illegible] United States currency at any time within five years from the 1st January [illegible]; the bonds to bear 8 per cent. interest per annum, payable half-yearly. On the 18th January, 1870 the Directors authorized the issue of these bonds in sums of not less than $1000 each. They are payable at the Bank of America in the City of New York. Of these bonds $187,810.99 were exchanged for the then outstanding bonds of the Company, the old bonds being cancelled and the mortgages securing them released. During 1870, a floating debt of $[illegible] was cleared off, the bond interest to the amount of $11,750.51 was also paid, along with other liabilities.

For the first two years the operations of the mining Company met with but little success, owing to unexpected and unavoidable mishaps at their mines. The work was, however, prosecuted with commendable perseverance and there is now a good prospect that all difficulties will be overcome. A vein of ore has been struck of superior quality, said indeed, to be equal to the best Lake Superior ore; of this large quantities are now being turned out. This, with the very considerable lumber traffic derived from the Company's Mills on the north shore of Rice Lake, gives the Railway as much as it can do; and it is hoped that the enterprise has entered at last upon a career of permanent prosperity.

The line of railway extends from Cobourg to Harwood, 14½ miles, with two sidings from Main Line into Campbell's and Macdougall's Steam Saw Mills, one mile each; and the unfinished line North of Rice Lake to Peterborough, which only requires new rails. A branch line, nine miles in length from the Narrows on the River Trent to Blairton, where are the Company's extensive and valuable Iron Ore Beds. Total 25½ miles.

The Company has other property consisting of twenty-three thousand acres of land in the townships of Belmont, Marmora, and Lake, including the Iron Mines, the village of Blairton, containing Railway Depot, Engine House, and 60 tenements built by the Company for the comfortable accommodation of the Miners and Employees. Also part of the village of Marmora, with water power, saw and grist mills, and buildings, besides sheds, store houses, &c.

The rolling stock consists of two first-class freight engines; two second-class freight engines; fifty platform cars; two second-class passenger cars; one hundred Iron Ore dumping cars. The steamer "Otonabee" is used to carry the ore across Lake Ontario and eight scows are employed for the transportation of ore and other freight on Rice Lake. There is a steam elevator for elevating ore at Harwood, and an elevated dock, erected by the Company at Cobourg for the purpose of dumping the ore direct from the cars into vessels.

RECEIPTS, 1869.—Received from sales of Iron Ore, $[illegible]; for carriage of lumber, $17,757.59; for carriage of grain and flour, $[illegible]; from miscellaneous freight, $2,[illegible]; received from rents, $[illegible]. Total, $[illegible].

EXPENDITURES, 1869.—*Permanent Improvements.* - Fifty iron ore dumping cars, $12,500.00 ; Elevated Dock at Cobourg, $3,000.00 ; Wharf Extension at Harwood, $600.00 ; Steam Elevator at Harwood, $500.00 ; Tanks and Telegraph Line, $500.00 ; McDougall's and Campbell's Sidings, $2,000.00. Total $19,100. *Operating Expenses.*—Mining Ore, $21,000.00 ; Operating Road, $10,000.00 ; Staff Salaries, $4,000 ; Lake Freights on Ore, $12,000.00 ; Duty on Ore, $4,800.00 ; Handling Ore, $2,400.00 ; Harbour Tolls on Ore, $1,200.00 ; Rice Lake Transportation, $3,100.00. Total, $58,500.00. *Interest.*—One year's Interest on Bonds, $10,984.00 ; Bank Interest, $2,780.00 ; Grand total $91,364.00.

MANAGING DIRECTOR.—W. P. Chambliss.

SUPERINTENDENT.—James R. Barber.

CHIEF OFFICE.—Cobourg.

EUROPEAN AND NORTH AMERICAN RAILWAY.

A line of Railway to connect St. John, on the Bay of Fundy, with Shediac, on the Gulf of St Lawrence, was first projected in 1848. In that year the sum of $4,000 was granted by the New Brunswick Legislature towards paying the expenses of a preliminary survey, which was made in the following season. During the next year a bill was passed by the House of Assembly of that Province, which provided that $600,000 stock should be subscribed by the Province and a like amount by private parties, and that the Province should guarantee six per cent interest on the sum of $12,000,000 in aid of the Railway. This measure failed to secure the sanction of the Legislative Council.

It was in 1850 that the project for the construction of the railway took tangible shape. In that year a Convention, composed of delegates from the State of Maine and the Provinces of Nova Scotia and New Brunswick met at Portland, Maine, for the purpose of discussing the proposal to construct a railway to connect Halifax with Bangor, Me. At this Convention, the scheme of the European and North American Railway was approved and decided upon. Exploratory surveys were made in the same year by authority of the State Legislature.

In 1851 the Act known as the Facility Bill was passed. This Act provided that a subsidy of $1,200,000 should be granted in aid of the enterprise, in the shape of debentures bearing six per cent interest, and redeemable in thirty years. As soon as $500,000 of capital was paid in by the subscribers to the stock, the Local Government were to issue their six per cent debentures to a like amount, the issue in one year not to exceed $200,000. The Board of Management was to consist of nine directors, two of whom were to be elected by ballot, (both Houses of the Provincial Legislature voting), to represent the Province.

A contract was entered into with Messrs. Peto, Betts, Jackson and Brassey, on the 29th September, 1852, by the Government of New Brunswick for the construction of the Road. By the terms of this contract, the contractors were to build the Railway from the boundary of Nova Scotia, to that of the State of Maine for $32,000 per mile. The Province was to take stock to the amount of $6,000 per mile, and to loan its bonds to the Company for $9,600 per mile. These were preference bonds and were redeemable in twenty years. At a special session of the Legislature called the following month, this contract was duly ratified.

In the following year (1853), surveys of the whole route were made in Nova Scotia and in New Brunswick; and on the 14th September, the first sod was turned by Lady Head at St John, N. B. Construction was immediately commenced between St. John and Shediac, and prosecuted during that and part of the following season, when in consequence of financial embarrassments growing out of the crisis that overtook these provinces in common with other countries after the close of the Crimean war put a stop to further operations.

The company of contractors was dissolved in 1856. The Government then purchased the road from them for the sum of $450,000, and continued the work under their own supervision. In the spring of 1857 the undertaking was placed in the control of three commissioners who held office only for a few months, when they were succeeded by three other gentlemen. From May 1858 till June 1865, this board consisted of R. Jardine, R. C. Scovill, and George Thomas.

Prior to the transfer from the first contractors to the Government, as before mentioned, the line had been located and surveyed from St John to Shediac. Between Moncton and Shediac a considerable portion was built, and some work was done on other parts of the line. On the 1st August, 1856, a contract was let for finishing the line between Moncton and Shediac; this section was completed on the next year, 1857. A short piece of three miles, out of St John, had been opened on the 17th March, 1857. As soon as a revision of the location could be completed, other sections were put under contract, completed and opened for traffic at the dates following:—St John to Rothesay, 9 miles, on 1st June, 1858; Rothesay to Hampton, 18 miles, on 8th June, 1859; Hampton to Sussex, 22 miles, on 10th November, 1859; and Sussex to Moncton, 45 miles, on the first August, 1860, thus completing the whole line from St John to Shediac, a distance of 108 miles.

There was nothing done in furtherance of the project until 1864, when it was again revived, and surveys were made under instructions from the New Brunswick Government from St John to the American boundary, and from Moncton to the Nova Scotia boundary; the former by Mr. Burpee, and the latter by Mr. Boyd. Meantime the Government of Nova Scotia had constructed the road from Halifax to Truro, and opened it for traffic. Two companies, one in Maine and the other in New Brunswick, were incorporated to construct the remaining portions of the line on both sides of the boundary respectively; subsidies were also granted by the legislatures of Nova Scotia and New Brunswick and by that of the State of Maine. At the commencement of the present year, the line to Sackville was formally opened from Moncton, a distance of 32 miles, and 123¾ from St John.

The line from St John to Point du Chêne, in Shediac harbour, is 108 miles; maximum gradient, 45 feet to the mile; minimum radius of curve, 1534 feet; the highest summit is 165 feet above high water in St John harbor; total length of straight line, 79½ miles; of curved line, 28¾ miles; it is a single track road of 5 feet 6 inches gauge; length of sidings, 12.9-10 miles. About 20 miles of rails were laid of the U pattern; the rest was the T rail, of 63 pounds to the yard, fastened at the joints with cast iron chairs, weighing 28 pounds each; the sleepers are 9 feet long, 6 inches thick, and of cedar, hackmatac and pine; width of road bed, 20 feet on embankments, and 30 to 32 feet in sidehill cuttings. There are 25 bridges having stone abutments and wooden superstructures, the remaining 8 are on piles.

The capital account was debited as follows, on the 30th June 1870: Cost of road and equipment $4,703,385.16; Balance to Dominion account $112,535.82 —Total $4,815,920.98. The cost of the road is made up of the following items:—Engineering $216,878.62; Permanent Way $3,648,658.38; Buildings $168,017.75; Rolling Stock and Machinery $407,834.64; Fencing $88,000; Wharves $93,433.15; Miscellaneous Stock $15,512.03; Miscellaneous $65,-055.59.—$4,703,385.16.

Working expenses in 1868-69, 69.01 per cent.; in 1869-70, 71.42 per cent. of the gross receipts; earnings per mile in 1868-69, $1,691.62, expenses $1,-168,05; earnings in 1869-70, $1,539.82, expenses $1,099.87.

The maximum per centage of useful load to dead weight carried was in 1864, 37.95 per cent.; in 1865, 35.19 per cent.; in 1866, 25.81 per cent; in 1867, 36.82 per cent.; in 1868, 36.77 per cent.; in 1869, 36.81 per cent.; in 1870, 36.02 per cent.

The Railway was transferred to the Dominion Government under the terms of Confederation, and is now operated on behalf of the Receiver-General of the Dominion, under the immediate superintendence of Lewis Carvell, Esq.

OFFICERS.—Lewis Carvell, General Superintendent; Alex. McNaughton, Accountant: James Coleman, Transportation Master; Gavin Rainne, Track Master.

CHIEF OFFICE—St. John, N. B.

WESTERN EXTENSION RAILWAY.

This Line is a part of the "European and North American" system, being the extension of that Railway from St John westward to the Boundary of the State of Maine. It is under the management of a Company having separate and distinct corporate powers, obtained from the Legislature of New Brunswick. The portion of the same road on the American side is under the management of another Company chartered by the Legislature of Maine. The Western Extension touches the boundary at Vanceboro, whence the line is continued to Bangor, Me. On the American side the road is open from Bangor to Mattawamkeag, a distance of 58 miles; the remainder of the line between Mattawamkeag and Vanceboro—the junction of the American and Canadian lines—is under construction, and is being pushed forward with vigor; it is expected that it will be completed by the 1st October next. When this connection is made a through line of travel and traffic will be opened between the railways of the United States and those of the Dominion in the Maritime Provinces.

Construction was commenced in August, 1867, and the Line was opened for traffic in August, 1869—showing that the enterprise was taken hold of by energetic hands, and pushed rapidly forward to completion.

Length of Line, 88 miles, sidings 2 miles—total, 90 miles. Weight of rail 56 lbs. to the yard; gauge 5ft. 6in. All the longer bridges have iron superstructure.

The capital stock was subscribed as follows:—

By the Government of New Brunswick	$300,000
By Individuals in United States	250,000
By Individuals in New Brunswick	193,750
By the City of Saint John	60,000
Total	$803,750

The amounts of the above subscriptions paid in are—

Government of New Brunswick	$300,000 00	
Individual subscribers (N. B.)	120,463 74	
City of St. John	60,000 00	
	$480,463 74	
Add unpaid	323,286 26	—$803,750

The $250,000 subscribed in the United States is held by the contractors for building the road; and it is reported that when they are settled with, that amount will be entirely absorbed. Difficulty has been experienced in realizing the subscriptions of private parties in New Brunswick, and nearly the whole amount unpaid is in litigation, the subscribers having repudiated their liability to pay. One of these cases was tried in January, 1871, resulting in favour of the Company; but it was appealed, and will be argued in June next.

The Company have authority to issue bonds to the amount of about $2,000,000. Of this amount $1,716,961 26 have been issued up to the 31st May, 1870. These were principally negociated in London, Eng., for iron and cash; they are in denominations of £200 stg. each, with interest coupons, payable on 1st January and July at the rate of six per cent per annum at the banking house of J. S. Morgan & Co, London, England.

Under an Act of the Legislature of New Brunswick, passed in 1864, the Company is entitled to a subsidy of $10,000 per mile, and as the length of the road is 88 miles this subsidy amounts to $880,000; of this amount $830,000 had been received up to January, 1871.

The road is in want of good actual sidings, the cars having to load, in many places, on the main track. This want will be supplied in part at least by the expenditure for that purpose of the sum of $10,000, which amount represents the *net profit on the first year's operations.*

The total expenditure on capital account up to the 3rd May, 1870, was $2,692,894 51.

Rolling Stock.—Number of Locomotives, 6; Passenger cars, 4; Freight cars, 115; Baggage cars, 2; Mail and Express, 2.

Mileage of Passenger cars (year ending Dec. 1st, 1870), 138,198; Freight cars, 215,856; Locomotives, 64,413; Passengers carried (number), 50,766; Freight, tons, carried, 33,329.

Revenue (year ended Dec. 1st, 1870).—Passenger traffic, $39,062 74; Freight, $26,515 65; Express, Mails and Sundries, $4,383 84. Gross earnings, $69,962 23.

Gross Expenditure—$59,831 61.

General Balances.—Capital stock, $808,750; Bonds issued, $1,716,961 26; Bonuses, $800,000; Other accounts, &c., &c., $108 83.

Per Contra.—Construction, $2,583,914 60; Rolling stock, $108,979 91; Stock, $323,286 26; Current assets, $304,639 32.

Directors (elected 15th June, 1870).—Alexander Jardine, Esq., Major W. B. Robinson, Lewis Carrell, Esq., Hon. Thos. R. Jones, and James R. Reed, Esq.

Officers.—Alexander Jardine, President; T. Barclay Robinson, Secretary-Treasurer; C. N. Skinner, Solicitor; E. R. Burpee, Manager; H. D. McLeod, Superintendent.

INTERCOLONIAL RAILWAY.

The project of a railway, connecting Quebec with the seaports of Halifax and St John, has been long cherished as a necessary connecting link between the British Provinces of North America. Though agitated at various times, the idea only took practicable shape when the present confederation was determined upon and arranged at Quebec, and by the 145th section of "The British North America Act, 1867," commonly called the Union Act, the construction of the railway was made obligatory upon the Government and Parliament of Canada.

A good deal of time and money have been spent in surveying different routes and examining the country through which the road is to pass. Three principal routes were surveyed, known as the "Frontier," the "Central," and the "Bay Chaleurs" routes. A table extracted from Mr. Sandford Fleming's report, shows the distances by the different surveys (fifteen in number), between River du Loup and St John and Halifax.

TABLE OF COMPARATIVE DISTANCES FROM RIVER DU LOUP TO ST. JOHN AND HALIFAX.

		TO ST. JOHN.			TO HALIFAX.		
ROUTES.	No. of line.	Railway Built.	Not Built.	Total.	Railway Built.	Not Built.	Total.
Frontier Routes.....	1	27	292	319	184	401	585
	2	45	306	350	202	414	567
	3	00	301	301	157	410	561
Central Routes....	4	00	326	326	157	435	592
	5	00	328	328	157	437	594
	6	37	343	380	120	452	572
	7	77	349	426	80	458	538
	8	37	307	344	120	416	536
	9	77	313	390	80	422	502
	10	96	326	422	61	435	496
	11	37	323	360	120	462	582
	12	77	329	406	80	436	516
Bay Chaleurs Routes...........	13	37	357	494	120	496	616
	14	96	377	473	61	486	547
	15	96	390	486	61	499	560

The route adopted is that known as the North Shore or Major Robinson's route, and is No. 15 of the above table. In compliance with addresses presented to the Imperial Government about 1845 by Nova Scotia and New Brunswick, the Imperial Secretary of State offered to have the Line surveyed by an officer of the Royal Engineers provided Nova Scotia and New Brunswick would share the expense. This offer was accepted and Major Robinson's report was the result. The proposed railway will therefore run from Halifax to Truro at the head of the Bay of Fundy, passing over the Cobequid Hills, and on and near to Amherst and Bay Verte, crossing from these over to the River Richibucto and Miramichi, then by the valley of the north-west Miramichi and Nipisguit River to Bathurst; then along the shore of the Bay Chaleurs to the Restigouche River; then by the valley of the Matapedia over the River Metis; then along the banks of the St. Lawrence, at a distance of eight or twelve miles from the south shore to Riviere du Loup. The distances to Halifax by this line are estimated as follows :—

	Miles constructed	Not constructed	Total.
From Riviere du Loup, by Metis, Matapedia, Dalhousie, and Bathurst to Muncton............	...	390	390
From Moncton to Truro............	...	109	109
From Truro by Railway to Halifax.	61	...	61
Total..................	61	409	560

Very different views seems to prevail as to the desirability of the different routes. It is admitted, however, that the objects arrived at by the construction of the Line were political as well as commercial; and in view of the decided stand taken by the Imperial Government, whose guarantee was asked and offered to facilitate the raising of the necessary funds, it is difficult to understand how any other route could have been chosen.

The Duke of Buckingham's despatch, dated 22nd July, 1868, is as follows: "I have received your Lordship's telegraphic message that the route by the Bay of Chaleur has been selected by the Canadian Government, as the one to connect Truro with Riviere du Loup, and thus complete the Intercolonial Railway. I understand three routes to have been under the consideration of the Government of Canada, namely: one crossing the St. John River, either at Woodstock or Fredericton; the second in a more central direction through New Brunswick, and the third following the line selected by Major Robinson in 1848. The route crossing the St. John River, either at Woodstock or Fredericton, is one to which the assent of Her Majesty's Government could not have been given; the objections on military grounds to any line on the south side of the St. John River are insuperable. One of the main advantages, sought in granting an Imperial guarantee for constructing the railway, would have been defeated if that line had been selected. The remaining lines were the central line, and that following the general course of the route surveyed by Major Robinson; and Her Majesty's Government have learned with much satisfaction, that the latter has been selected by the Canadian Gov-

comment. The communication which this line affords with the Gulf of St. Lawrence at various points, and its remoteness from the American frontier, are conclusive considerations in its favor, and there can be no doubt that it is the only one which provides for the national objects involved in the undertaking."

On 19th April, 1867, an Act was passed by the Imperial Parliament authorizing the Commissioners of Her Majesty's Treasury to guarantee a loan not exceeding Three Millions Pounds Sterling, at a rate not exceeding four per centum per annum, to assist in the construction of the Railway, and providing that the guarantee should not be given unless and until the Parliament of Canada should, within two years of Confederation, pass an Act providing to the satisfaction of one of Her Majesty's principal Secretaries of State, as follows, viz.:—

I. For the construction of the Railway.

II. For the use of the Railway at all times for Her Majesty's military and other service.

III. Nor unless and until the line on which the Railway is to be constructed, has been approved by one of Her Majesty's principal Secretaries of State.

On 21st December, 1867, an Act was passed by the Parliament of Canada for the construction of the Intercolonial Railway. The Minister of Finance then placed a loan of Two Million Pounds Sterling upon the London market, seventy-five per cent thereof having the Imperial guarantee, and twenty-five per cent being without it; and the whole was taken up at once on favorable terms.

On 11th December, 1868, in terms of the Intercolonial Act, four Commissioners were appointed to construct the Railway. The Board consists of Aquila Walsh, Esq., M. P., North Norfolk, Chairman; the Hon. Edward Barron Chandler, member of the Legislative Council of New Brunswick; Charles John Brydges, Esq., Managing Director of the Grand Trunk Railway, and the Hon. Archibald Woodbury Whelan, Senator.

The whole length of Railway from Rivière du Loup to Truro, (including eight miles of the European and North American Railway and the Eastern Extension Railway thirty-seven and a quarter miles) is four hundred and ninety-nine and a half (499½) miles

The Railway (which is being constructed under the superintendence of Sandford Fleming, Chief Engineer) has been let in sections, and all the work is now under contract. These contracts include clearing, grading, fencing, and bridging, except in the case of the bridges over the rivers at Trois Pistoles, Metis, Restigouche, Nepisiqui, the two branches of the Miramichi, and Folly River. The bridges are all to be of wood, except at the places named, and the contracts do not include the iron superstructures at these places. The entire line is to be laid with steel rails.

The aggregate amount of the contracts for the whole line, including purchase money of the Eastern Extension Railway, is $10,513,781.

The line is completed from the European and North American Railway to

Amherst; and rails will be laid upon a number of the more advanced sections during the season of 1871. The rolling stock is being prepared by various parties who have contracted.

The Intercolonial Railway will connect with the Grand Trunk at Riviere du Loup; and being of the same gauge as the Grand Trunk (5ft. 6in.) freight and passengers can be sent from one end of the Dominion to the other without transhipment so soon as the Railway shall be completed.

NEW BRUNSWICK AND CANADA RAILWAY.

A number of inhabitants of the town of St. Andrew's in the County of Charlotte, Province of New Brunswick, convened on the 5th day of October, 1835, and formed an association under the appellation of the "Saint Andrews and Quebec Railroad Association" for the purpose of promoting the interests of a railway from the town of St. Andrews on the sea coast to the City of Quebec in Lower Canada, a distance of 195 miles. The former town was intended to be a winter port for the trade of the St. Lawrence.

The estimated cost of the road at that time, by making use of the flat rail which was then in use in the United States, was £5,000 per mile.

A deputation of two gentlemen was sent by the Association to the British Government in January, 1836, seeking aid, and they succeeded in obtaining a grant of £10,000 from His Majesty King William the Fourth, to be expended in a thorough exploration and survey (through a wilderness), which was commenced in June, 1836. On the 27th August the sum of £2,000 was received from England and deposited in the Charlotte County bank, this being the first instalment of the £10,000 grant. About the same time the Secretary of the Association received a letter from the Government prohibiting further explorations, owing to a representation from the United States, until the question of the north eastern boundary between Maine and New Brunswick should be settled.

Further proceedings on the part of the Association were now held in abeyance and remained so until 1845, that memorable period of the great railway and commercial panic throughout England when the speculative "King Hudson" was approaching the zenith of his popularity. It was during this period that the "Great Northern American Railway." was projected to connect Halifax and Quebec for the purpose of carrying troops and mails, but this scheme did not meet with success. The British Government expended the sum of £12,000 in explorations on this route between those cities.

The eastern boundary of Maine was settled by the Ashburton treaty in 1842, and the Association again revived. In the month of December 1845, a subscription list was opened. The capital stock of the Company was divided into 30,000 shares of £25 each. Over £41,000 was subscribed in the County of Charlotte. The Directors decided not to commence operations until £100,000 stock was taken.

On the 17th March, 1847, at a meeting of Directors, Captain J. Robinson, R. N., and Moses H. Perley of St. John, were appointed agents and sent to England to dispose of shares and to effect a loan. These gentlemen succeeded in disposing of a number of shares and formed a Board of management. An agreement was entered into and signed by representatives of the Provincial and English shareholders on the 15th July 1847, and the capital stock of th Company was divided into two classes, "Class A," consisting of 4,000 shares, to parties not being on the Continent of North America, and "Class B," consisting also of 4,000 shares, belonging to parties resident in New Brunswick, or elsewhere on the Continent of North America. Scrip certificates to be issued to Class A shareholders on payment of the deposit of £2 sterling per share, being equal to £2 10s. per share currency. Interest after the rate of £5 per cent, per annum, to be paid by the Company to the proprietors of shares in Class A until the completion of the road from St. Andrews to Woodstock. During the first ten years after completion, the clear profits arising from the traffic of merchandise and passengers, together with such sums of money as might be received from the Legislature of New Brunswick, to be applied in paying to the Class A and B shareholders a dividend of five per cent. on the capital subscribed for by them; and any residue capital to be divided amongst all the shareholders proportionately; the Class A in all cases not to receive less than five per cent. dividend for a period of ten years after opening of road for traffic. After the expiration of this period, the clear profits arising from the road to be divided amongst all the shareholders without any preference to Class A.

The Company to be represented by twenty Directors, thirteen in New Brunswick, and seven in England. No call should be made on the Class A shareholders before 1st day of January, 1848. In the event of the 4,000 Class A shares not being fully subscribed for before this period, such of the Class A who desired it would have their deposit money returned without any deductions, but without any interest, and shall have no further interest whatever in the Company.

The estimated cost of building the line from St Andrews to Woodstock was £160,000. An estimate was also made of the probable earning and expenditure and a net profit being equal to 20 per cent on the capital was the result.

At a meeting of Directors held the 21st August 1847, it was decided to commence operations as soon as possible and to engage a competent engineer at once.

On the 28th October another meeting was held and tenders for masonry and bridging the first 4 miles were invited. The ground was first broken in rear of the town of St. Andrews in November of the same year, and the work commenced by day laborers.

During the month of March a Committee of the Legislature recommended an issue Provincial Scrip to the extent of £50,000, redeemable in forty years, upon the stock amounting to £150,000 being paid in. The Province also to guarantee sixper cent. for 25 years on £100,000. A grant of 40,000 acres of land in alternate blocks, was also recommended. On the 24th of April the Legislature and Council passed the Facility Bill granting the above guarantee and 20,000 acres in alternate blocks.

Owing to delays and difficulties operations were almost entirely suspended, and in February 1851, the Company again advertised for tenders On the 15th April a contract with the Messrs. Myers was signed for grading the first ten miles from St. Andrews.

The Board of Directors made a conditional contract with the Messrs. Myers on the 18th June to build the entire line to Woodstock in eighteen months from date, for the sum of ten thousand dollars per mile. This included grading, bridging, laying ties, and rails, the Company delivering the latter at St. Andrews, and all the necessary rolling stock required by the contractors. During the month of August the first grant of 10,000 acres from the Government was conferred upon the Company and conveyed to the Managing Director at London for the benefit of Class A shareholders.

On March the 11th the first cargo of rails and a locomotive and tender named "the Pioneer" arrived from Newport, and during the month of April a second cargo of rails arrived. At the time this cargo arrived the Company had not sufficient means to pay the freight and were obliged to offer a mortgage payable in two months, which was agreed to.

His Excellency the Lieut.-Governor appointed a Committee of the Attorney and Solicitor General to investigate the accounts of the Company. Their statement was dated 5th November, 1851, which showed a deficiency of capital amounting to £101,294.

The total expenditure amounted to £35,614, about 22 per cent. of which amount was absorbed by the two Directories law and interest account. The liabilities of the Company at this date were £5,435.

There still remained 70 miles to build, estimated at £2,300 per mile	£161,000	
Rolling Stock and stations estimated at	10,000	
Adding 20 per cent. for contingent expenses in England and the Province, Law, Interest, &c	34,200	
Total Sterling	£205.200	

Resources of the Company were

Balance due on Class A Stock	£66,907	
" " Class B "	36,886	
Grant of Lands 20,000 acres at 10s	10,000	
		113,793
Leaving a deficiency of		£94,407

The grant of lands was not available unless a sufficient expenditure was proved by the Company to entitle them thereto; and therefore such a resource was entirely conditional upon stock being paid up.

Notwithstanding this state of affairs the Company seem to have been determined to struggle for the completion of the work. In the autumn of 1851, they made a contract for the completion of the line, but the security being unsatisfactory the work was not performed.

In March 1852 the London Board made an advance of £4,000 sterling to the Company to meet their liabilities in the Province, and, two months after, the Company entered into a contract with Messrs. Sykes & Co., to build the road from the end of the first 10 mile section to Woodstock, a distance of seventy miles, for the sum of £161,000 sterling. On the 4th of June the turning of the "first sod" on the new contract was celebrated, and in a short time the work was commenced and pushed vigorously. Money was coming in freely and the affairs of the Company seemed to be in a more prosperous condition, but this happy state of affairs did not last very long. In Sept. 1853, the local and English Board of Directors who had acted harmoniously up to this date unfortunately had some misunderstanding, and a Mr. Byrne was sent from England to arrange matters then in dispute. The breeze between the two Boards had scarcely subsided into a calm, when signs of a threatening squall arose between the local Board and the contractors. The latter demanded an arbitration on disputed claims which was waived by the payment of £1,500, but this did not seem to satisfy the Contractors. The Board then served a notice preparatory to making an entry under the contract. At this time there were only about 25 miles finished. After the Company came into possession of the line the next difficulty was how to proceed with the work and furnish the means for so doing. Various Board meetings were held up to January 10th, 1855. At this date Mr. Byrne on behalf of Class A proposed to take the whole charge and control of the road to Woodstock, but this it seems was not accepted. Matters again apparently came almost to a dead lock. Frequent meetings of the Board were held with but little progress except getting several Acts passed by the Legislature to enable the Company to transfer their stock and get an extension of time.

At a general meeting of stockholders on the 6th May, 1856, after the election of Directors, it was resolved to authorize a transfer of the corporate powers, &c., of the Company to Class A shareholders, or to a Company formed in accordance with a scheme agreed to by the Class A Board of Directors and also those of the Class B.

About this time a new Company was formed in England for the completion of the railway to Woodstock, and at a meeting held on the 16th October the representatives of the three different Companies were present and a deed of transfer was made to the new Company. Mr. Julius Thompson was appointed manager. The new Company commenced operations at once by letting contracts for repairs, ballasting, &c.

On the 1st October 1857, the road was opened to a distance of 34 miles, and in December 1858, to Canterbury, a distance of 66 miles. Mr. Thompson was succeeded as manager by Mr. Henry Osburn, who concluded a contract for the completion of the line to the Richmond terminus on the main road between Woodstock and Houlton, which was opened for traffic in July, 1862. The contractors were paid in first mortgage bonds, at 20 per cent. discount, bearing 6 per cent interest. On account of the great difficulty experienced in floating these bonds the Company were obliged to suspend operations, but by temporary bridging in lieu of embankment they succeeded in getting the rails laid to the terminus.

In consequence of the inability of the Company in England to meet the amount of interest on the Mortgage bonds, the Manager, Mr. Osburn, was also appointed Receiver in 1863, (and still holds this position) on the part of the bondholders, and the line has since been worked for their benefit ; but as the Company then owned so small a quantity of rolling stock, and the line being left unfinished under the last contract, it became necessary to expend from year to year, in addition to the cost of maintenance, large sums out of the earnings in completing the earthworks, masonry and ballasting, and for increasing the rolling stock and machinery, for repairs to locomotives, &c., the balance of excess being held by the bond holders. Since the opening of the main line, two branch lines have been opened and are now run over—one from Saint Stephen, 19 miles in length, opened January, 1866, the other from Woodstock, 11 miles, opened in September, 1868. These branches were built by provincial Companies facilitated by the Local Government Subsidy Act, passed 11th April, 1864, which granted a bonus of $10,000 per mile, in aid of the construction of proposed railways therein mentioned.

	1864.	1865.	1866.	1867.	1868.	1869.	1870.
No. of Locomotives	5	5	6	6	10	10	....
" Pas'ger cars.	4	4	5	5	8	9	....
" Freight "	36	53	62	63	68	76	....
" other "	4	6	6	6	30	30	....
Mileage of Locomotives................	68,024	77,662	85,358	73,809	88,407	115,822	....
Passengers carried (number)..........	6,431	8,038	8,243	15,550	16,501	23,503	....
Freight, tons, upward................	3,911	4,222	6,461	6,520	8,157	8,900	... .
Freight, tons, downward................	27,179	37,347	46,102	49,686	60,414	59,405	... .
REVENUE.	1864.	1865.	1866.	1867.	1868.	1869.	1870.
Passenger Traffic...	6,968	7,805	8,610	12,894	16,379	19,864	29,120
Freight " ...	40,017	45,433	56,287	66,389	86,672	84,364	111,478
Mail and Sundries.	182	666	425	498	297	600	842
Gross Earnings	47,167	53,904	65,322	79,781	103,348	104,828	141,440
EXPENDITURE. Operating	36,803	38,619	45,646	53,506	63,116	75,926	....

Permanent Way.—Length of Line, [illegible] miles; Length of Branches, [illegible] miles; Length of sidings, 12 miles. Total length, [illegible] miles. Weight of Rail per yard, 56 lbs.; Gauge, 5 feet 6 inches; Termini of Main Line, St. Andrews and Richmond; Termini of Branches, St. Stephen, Woodstock, N. B., and Houlton. The amount expended on construction up to 18[illegible], $[illegible].

OFFICERS.—Henry Osburn, C. E., General Manager; J. P. [illegible], Superintendent; N. T. Greathead, Cashier; A. E. Julian, Ticket Agent.

CHIEF OFFICE—St. Andrews, N. B.

WINDSOR & ANNAPOLIS RAILWAY.

During the year 1864 the Nova Scotia Government propounded a new policy for the extension of this line from Windsor to Annapolis. The latter is a small town on the Bay of Fundy, which was once the capital of British North America, and was settled in 16[illegible] by the French. The features of this policy may be stated thus:

1st. The right of way valued at £60,000 or £70,000 was granted by the Counties through which it passes to the Company, with the privilege of possession as required, regardless of indemnity. A special tax to be levied on the Counties for the payment of the same.

2nd. The free use of timber and stone on the Government lands.

3rd. The free use of the Government Railway and wharf at Halifax for the transport of all material supplies, &c., the Company being only at the expense of handling.

4th. Rebate of all duties, imposts, &c., on material used in construction.

5th. The sum of £32,000 in cash to make the construction of the bridge over the Avon. A bonus of £133,600 in 6 per cent. bonds, payable as the work progresses. These items and subventions amount to over £3,[illegible] per mile as an actual gift and totally irrespective of the receipts or ownership of the railway which are for the sole benefit of the Company.

On the above basis a Company was formed, and on the [illegible] October, 18[illegible], a contract between the Chief Commissioner of Railways and Edward Harris and F. T. V. Smith, on behalf of Messrs. Knight & Co., of London, England, the work to be commenced by the first of May following, and the road to be completed and ready for traffic on the first May, 18[illegible]. This agreement was confirmed by George Knight & Co., but they failed to commence the work and the agreement was cancelled.

On the 22nd of November, 1866, Messrs. Tupper, Henry & Ritchie, then in England, having been authorized by an order in Council, and acting on behalf of the Chief Commissioner of Railways, entered into an agreement with Messrs. Punchard, Barry & Clarke, by which the latter were to construct the works which were to become their property, the work to be commenced not later than the 1st January, 1867, and to be fully completed on or before the first of Dec., 1869.

This line passes through the Annapolis valley, which is one of the oldest settled and richest parts of the Province, connecting with the Nova Scotia Railway at Windsor, 45 miles from Halifax, and at Annapolis with a line of steamers to St. John, New Brunswick, a distance of about 60 miles, making a total distance between Halifax and St. John of 190 miles.

The road was partially opened on the 11th August, 1869, and completed on the 18th of December of the same year. During the first six months the line was by agreement worked for the benefit of the contractors. The length of the main line is 84 miles, with 8 miles of sidings. The gauge is 5 feet 6 inches. The rails are fish-jointed, and between Windsor and Kentville they weigh 67 lbs per yard, and the remainder of the distance they are lighter, weighing only 52 lbs. per yard. The rolling stock is substantially constructed and consists of nine locomotives, twelve passenger and 120 other cars. The most important feature of the Line is the iron bridge over the Avon at Windsor, where the tide rises over 40 feet. The bridge rests on stone piers. There are nine spans of lattice, iron girders. The total length is 1,130 feet. The total amount expended on construction account amounted to £542,332 sterling on the 30th June, 1870.

Directors.—George Sheward, Lord Allan Churchill, Colonel Cole, Albert Ruardo, Francis Lothell, John A. Bastard.

Secretary.—C. A. Tallbot, Westminster Chambers, Victoria Street, London.

General Manager.—Vernon Smith, Kentville, Nova Scotia.

Chief Office—Kentville, N. S.

CANADA SOUTHERN RAILWAY.

(PROJECTED.)

This line is located through the southern ten Counties of the Province of Ontario, an exceedingly prosperous agricultural district. Its eastern terminus is at the Intercolonial Bridge now being constructed over the Niagara River at Fort Erie, and its western termini are at Amherstburg, on the Detroit River, and at Moore, by a branch line of 60 miles, on the St. Clair River. The distance from Fort Erie to Amherstburg is 229 miles, and to

Thence 150 miles. It is intended to connect with the Michigan Southern and the Michigan Air Line on the west, and with the New York Central and Erie Railways on the east. The total length of line to be constructed is 290 miles. It is claimed by the promoters of this road that it is the shortest route from Buffalo to Chicago by 20 miles, as compared with the Canada Air Line, which is shorter again by 12 to 20 miles than any other route. Great stress is laid by the promoters upon the fact that 96 per cent. of the line is straight, with no opposing grade exceeding 15 feet per mile.

Mr. Courtwright, the President, states the financial scheme in this way :—

ASSETS.

Capital stock		$10,000,000
Capital stock subscribed		2,000,000
Leaving unsubscribed		$8,000,000
First mortgage, 7 per cent., sinking fund bonds		$9,500,000
Bonuses from municipalities		500,000
Cost of the road and equipment	14,500,000	
Commissions, expenses and contingencies	1,500,000	
		$16,000,000
He estimates the annual gross receipts of the undertaking at		$5,000,000
The working expenses, at 55 per cent		2,750,000
Leaving net		$2,250,000

From which, deducting interest on the bonds, ($712,000), leaves a net estimated revenue of $1,538,000. The subscribers to the $2,000,000 capital are offered the option, and declare their intention to take six millions additional of the capital stock, the bonuses, the proceeds of the two million dollars subscribed, and $8,000,000 of bonds, and furnish the necessary means to carry out the undertaking. It is the design of the promoters to have the road completed by the 1st of January, 1873.

It is to be built and equipped in first-class manner throughout. The track will be of the best Bessemer steel rails, weighing 60 pounds to the yard, with 2,500 cross ties to the mile, laid with chairs and fish-joint, one upon gravel ballast. The gauge is the American standard, viz.: 4 feet 8½ inches. Grading was begun last October on all heavy work, and is progressing rapidly. The bridging is all under contract and the timbers mostly on the ground.

The municipalities that have voted bonuses to this enterprise as follows :—Elgin, $200,000; Kent, $60,000; St. Thomas, $25,000; Amherstburg, $15,000; Anderdon, $15,000; Townsend, $30,000; Dunham, $15,000; Norwich, $15,000—making altogether close upon $400,000. Bonuses have been asked from other municipalities.

Trustees of the municipal bonuses have been appointed are as follow :—Hon. Wm. McDougall, A. McKellar, M.P.P., and Hon. H. Killaly.

DIRECTORS—Milton Courtright, Erie, Pa.; John F. Tracy, Chicago; Daniel Drew, New York; Sidney Dillon, New York; William L. Scott, Erie, Pa.; Wm. A. Thompson, Queenston, Ont.; John Ross, New York; O. S. Chapman, Boston, Mass.; Benjamin F. Hain, New York.

OFFICERS—Milton Courtwright, President, Erie, Pa.; N. Kingsmill, Secretary, Toronto; M. H. Taylor, Treasurer, Fort Erie; F. N. Finney, Chief Engineer, Fort Erie; William J. McAlpine, Consulting Engineer; Crooks, Kingsmill & Cattenach, Solicitors, Toronto; Charles J. Tracy, Solicitor, New York.

CHIEF OFFICE—Fort Erie, Ont.

LONDON, HURON & BRUCE RAILWAY.

(PROJECTED.)

It is proposed to build a line from the city of London, Ont., to some point on Lake Huron, most probably Goderich or Kincardine. The distance to the latter place is stated at 105 miles, and the estimated cost of the road is figured up at $800,000 to $1,000,000. It would pass through or near the Townships of London, Biddulph, McGillvray, Hay, Osborne, between Stanley and Tuckersmith, through Clinton, Wawanosh, and the village of Lucknow, Ashfield, Bruce, Huron and Kinloss townships, to Lake Huron. It is expected that London will give $100,000, a By-law having been introduced to that effect, and the townships along the line will also give bonuses to the amount of $250-,000 to $300,000. The Ontario Government will extend aid to the project to the extent, most probably, of $3,000 to $4,000 per mile.

CHIEF OFFICE—London, Ont.

TORONTO & MUSKOKA RAILWAY.

(PROJECTED.)

This line will extend from Barrie, where a junction will be formed with the Northern Railway, by the villages of Orillia and Atherly, and via Washago and Gravenhurst, to Bracebridge, in the Muskoka District. The route has been surveyed and located, and the right of way mostly secured. A contract for construction has been let to Messrs. John Ginty & Co., for the lump sum of

$[illegible]. The bridges are to be of stone and iron, and culverts of stone throughout.

The enterprise has been aided by a grant of $[illegible] by the City of Toronto, $[illegible] by the town of Barrie, and some further amounts were obtained from Municipalities along the route. A lease to the Northern Railway Company of the line has been arranged, the chief provisions of which are: 1st. That it should be for a period of 21 years. 2nd. That the Muskoka line should be constructed upon a specific standard. 3rd. That the tariff of the Northern should at all times apply by mileage rates to the Muskoka traffic, save and except that special provisions were made for cordwood. 4th. That the Northern should provide the necessary equipment in rolling stock. 5th. That the Northern should guarantee the Muskoka debentures to a limit of $[illegible],000 per mile of railway. 6th. That in consideration of stocking and working the line, the gross receipts thereof should be divided between the two companies, as follows:—First five years, 75 per cent. to Northern and 25 to Muskoka; second five years, 60 per cent. to Northern and 40 to Muskoka; remainder of term, 55 per cent. to Northern and 45 per cent. to Muskoka; and, finally, that any new and additional works required on the Muskoka line to meet increase of traffic during the term should be provided by the Northern at 6 per cent. for the outlay.

DIRECTORS, elected 1870—Messrs. Frank Smith, Anson P. Dodge, John Turner, Robert Spratt, Robert Wilkes, W. H. Howland, S. B. Harman, N. Barnhart, all of Toronto; W. D. Ardagh, Barrie.

OFFICERS—Frank Smith, Esq., President, Toronto; F. W. [illegible], Secretary.

CHIEF OFFICE—Merchants' Exchange, Toronto.

NORTH GREY RAILWAY.

(PROJECTED.)

This line is to extend from Collingwood, the present terminus of the Northern Railway, to Meaford, a village 22 miles distant westward, and situated on the shores of the Georgian Bay. The road is estimated to cost $11,000 to $12,000 per mile, or in all $[illegible]. Of this amount, the municipalities have voted one-half in the shape of bonuses, thus:—St. Vincent, $[illegible]; Collingwood, $[illegible]; Euphrasia, $[illegible]; total, $[illegible]. The Northern Railway Company is to take these bonuses and construct the road. A company has been organized, consisting of the reeves and others of the municipalities interested, to guard their interests in the undertaking. It is arranged that the line will be stocked and worked by the Northern Railway Company, on the same terms and conditions substantially as in the case of the Muskoka Railway. (See Toronto and Muskoka Railway.)

It is thought that the road will be in operation inside of a twelve-month. The gauge will be 5 feet 6 inches, same as that of the Northern.

DIRECTORS—(Elected April, 1871,)—Noah Barnhart, Toronto; C. R. Sing, Meaford; James Stewart, Meaford; T. Andrews, Thornbury; —— Noake, Euphrasia; —— Heward, Toronto; F. W. Coate, Toronto; H. L. Hime, Toronto.

OFFICERS—Noah Barnhart, Esq., President, Toronto; C. R. Sing, Esq., Vice-President, Meaford, Ont.

NORTH SHORE RAILWAY.

(PROJECTED.)

This project has excited a great deal of public interest in the Province of Quebec during the past year. The road is to run from Quebec to Montreal, on the north side of the River St. Lawrence, a distance of 190 miles. It is intended to effect a junction with the Northern Colonization Railway at the east end of Montreal. The project was set on foot in Quebec, and chiefly promoted in the interests of that city. At the last session of the Quebec Legislature an Act was passed authorizing the Governor-in-Council to aid the construction of this line, the building of a wooden railway from Three Rivers to Les Grandes Piles, (known as the Piles Railway), and to the establishment of a line of steamers on the St. Maurice River, by a grant of land to the North Shore Railway and St. Maurice Navigation and Land Company, to the extent of 3,203,500 acres. These lands lie in the counties of Pontiac, Portneuf, Quebec, Montcalm, Champlain and Chicoutimi, in four separate blocks. These lands were granted on the following conditions:—(1) That the Government shall not be bound to grant the lands until the road is complete and in operation, and steam navigation commenced on the St. Maurice. (2) That the Government may, at any time after 25 miles of the road is completed, make over to the Company a proportionate share of the grant. (3) That the Government appoint one-third of the Directors of the Company, without taking into account Directors *ex officio*, and that no municipality shall have a larger representation at the Board than the Government.

At a meeting held in Quebec, the President of the Company, Hon. M. Cauchon, stated their available resources, thus:—Two Parliamentary grants of land—one of 1,200,000 acres in the Ottawa Valley, another of 1,500,000 acres in the St. Maurice district, a vote of $1,[illegible] from, and the right of way in, the city of Quebec, as well as subscriptions to the amount of $[illegible] from the municipalities on the route between Quebec and Montreal. They had consequently $1,[illegible] in money and 2,700,000 acres to begin operations with, but it is necessary, in order to secure the land grants, that the road should be completed before the 1st July, 1875. The lands in the St. Maurice Valley are estimated to be worth $1 per acre, and those in the Ottawa Valley $2 per acre, making the value of these lands $4,400,000. The cost of the road at $30,000 per mile would be $5,760,000, which, after deducting the value of the lands and the bonuses, would leave the sum of $1,[illegible] to be raised in order to complete the road.

PRESIDENT—Hon. M. Cauchon, Quebec.

CHIEF OFFICE—Quebec.

WHITBY AND PORT PERRY RAILWAY.

(IN PROGRESS.)

This line is to run from Port Whitby, on Lake Ontario, to Port Perry, on Lake Scugog, connecting the inland waters of the counties of Ontario, Victoria, and Peterboro', for the purposes of trade with Lake Ontario. The canal lock at Lindsay being re-built, Lake Scugog, Mud Lake, Pigeon Lake and Chemung Lake, form a long line of water communication, on the borders of which a valuable and extensive lumber and milling trade is carried on. At present, this region finds an outlet at Port Hope, and lake cities further east. The promoters of this railway hope to control a considerable portion of this trade. The principal traffic would therefore consist of sawed lumber, square timber, cordwood, tan bark, shingles, grain and flour. The annual amount of this outward traffic has been estimated as follows:—30,000,000 feet sawn lumber, at $1 per M.; 15,000 pieces of square timber, at $1; 5,000 cords wood, at $1; 2,000 cords tan bark, at $1; 2,000,000 flour barrel staves, at 50c. per M.; 3,000,000 shingles, at 50c. per M.; 300,000 bushels grain, at 3c.; 10,000 barrels flour, at 10c.; 12,000 passengers, at 60c.; besides inward traffic, mails and sundries.

The assets of the Company on the 31st December, 1870, were as follow:—

Bonus Bonds of the Townships of Reach, Whitby, and the Town of Whitby, deposited with John Crawford, Esq., Toronto...	12,369 75
Bonus Bonds to be received from the Townships of Reach, Whitby, Scugog, and the Town of Whitby..........	42,000 00
Instalments due on Stock Calls to date..........	32,763 22

Balance due on Stock subscribed and not yet called in	31,155 00
Other Bonuses, not included in above	21,800 00
Interest on Town and Township Bonuses	5,740 00
	$181,852 97

The authorized capital is $300,000, and the subscribed capital $103,850. The receipts and payments for 1870 were:—

RECEIPTS.

Instalments on Stock	$ 33,966 92
Bonus Debentures from the Town of Whitby	30,000 00
Do. do. Township of Reach	20,000 00
Do. do. Township of Whitby	5,000 00
Interest collected on above Bonuses	3,267 50
Bills payable	1,105 90
First Mortgage Bonds	63,000 00
From other sources	2,786 05
	$159,125 47

PAYMENTS.

J. H. Dumble, account Contract	$101,800 00
For Right of Way	7,744 00
For Building Port Perry Dock	1,333 19
John Crawford, Esq., Toronto, Bonus Bonds and int. deposited	42,389 75
Preliminary Expenses	3,476 48
Office Expenses, Postage and Telegraph account	133 85
Taxes for 1870	9 56
Sundry accounts	2,198 91
Balance in R. C. Bank	34 37
	$159,125 47

The line is all graded; the ties are purchased and mostly distributed along the track. The iron for the permanent way has been obtained from the Aberdaire Company of Wales, and shipped so as to arrive about the 15th May. Some changes will be made in the road bed, by considerably reducing the steepest gradients; the intention being to make the road first-class in every respect. By the terms of the contract, the line is to be completed by the 1st August, 1871, but the contractors hope to finish a month sooner. The Company are building large wharves at Port Perry, where they will shortly erect an elevator. They are also arranging for wharf accommodation at Whitby, for the railway and traffic connected therewith.

DIRECTORS—(Elected 25th January, 1871,)—Joseph Bigelow, James Dryden, James Holden, N. G. Reynolds, Chester Draper, A. Ross, K. F. Lockhart, Thomas Paxton, M.P.P.; Edward Major.

OFFICERS—Chester Draper, President; Joseph Bigelow, Vice-President; Ross Johnston, Secretary.

TRUSTEE OF MUNICIPAL BONUSES—John Crawford, Toronto.

CHIEF OFFICE—Whitby, Ontario.

RICHELIEU, DRUMMOND, & ARTHABASKA RAILWAY.

(UNDER CONSTRUCTION.)

This line is 66 miles in length, and is to run from Sorel (P. Q.) to Acton, on the Grand Trunk Railway, passing through Drummondville, Yamaska, &c. It is stated that the intention is to build this line after the model of the Quebec & Gosford Railway, the rails being of wood. A contract has been let to Mr. Hulbert, the contractor of the Quebec & Gosford, and one or more locomotives and a number of platform cars have been ordered to be delivered early this season. The work of getting out ties and building bridges is now in progress.

BROCKVILLE & OTTAWA RAILWAY.

By this Company's Charter power was conferred to build a railway from the town of Brockville, on the River St. Lawrence, to the village of Pembroke, on the Ottawa River, with a branch from Smith's Falls—where the road intersects the Rideau Canal—to the town of Perth. The distance from Brockville to Pembroke is 130 miles, and from Smith's Falls to Perth, 12 miles. The line has only been opened to Sand Point, on the Ottawa River. The branch has also been completed, giving a whole length of railway of 90 miles.

Money was borrowed from the Municipal Loan Fund to aid the construction of the road as follows: Counties of Lanark and Renfrew, $800,000; town of Brockville, $414,491.96; township of Elizabethtown, 150,709.50—total, $1,366,201.46. The extent of these grants was a pretty good indication of the extravagant ideas that prevailed during the first Canadian railway era. The original expectation seems to have been that the profits these municipalities would derive out of the earnings of the railway would suffice to extinguish their indebtedness to the Government. This palpable delusion was soon dispelled. The road, as far as constructed, became deeply involved, and there were no funds remaining to complete the line to the Ottawa River, from which a large share of the traffic was expected. The position of affairs in 1862 and 1863 is thus depicted in the Directors' Report: "As this railway then stood—twenty-five miles short of its river terminus,- half-stocked, destitute of machine shops, and therefore working at the maximum of expense—the question when it would become a dead loss to every *bona fide* interest concerned rested solely upon the time when rails, engines, &c., should wear out, and heavy renewals become imperative." And further on they say, "such renewals could not have been adequately met from the limited income which it had power to earn, and to suppose that any interest to municipalities or bondholders could ever have been paid is simply preposterous." The traffic receipts were absorbed in payments of interest, so that the whole undertaking was on the high road to utter insolvency and complete ruin. In 1863 an Act was passed for the relief of the Company, which, though it was productive of good, did not prove sufficient to meet the exigencies of the case. By that Act the Company were authorized to issue preference bonds to the amount of $244,793.94, bearing 7 per cent. interest, for the purpose of extending the line to Sand Point, on the Ottawa, and that such should be a prior lien on the earnings of the road to the claims of the municipalities, and that the railway should repay the municipalities within fifteen years the sums paid by them to the Government under what was called "The Five per cent. Act" of 31st December, 1866, and to fund into 2nd class bonds the entire floating debt, principal and interest. The amount of this floating debt seems then to have been $711,019.97, besides $100,000 of unpaid interest due to the municipalities. That this measure was inadequate to relieve the road from its embarrassments is apparent from the fact that two years later—in 1865—the Company owed on preference bonds $244,793.94 ; 2nd class bonds, $1,098,285.77 ; unpaid interest, $150,000—total, $1,486,079.71. And the Company's whole liabilities, as harged to the debit of capital account, were $3,157,234.46, with credits of only $2,632,042.44, showing a deficit of $525,192. The earnings proved entirely disproportionate to meet the prior municipal and preference claims and the interest on the 2nd class bonds, so that it became apparent that further relief would have to be afforded, and the only shape that relief could take, in order to be effective, would be a liberal extinguishment of the debts, and the conversion of the remainder into stock.

A mortgage was made to a trustee to secure the re-payment of the preferential extension bonds of $244,793.94, above referred to. Owing to default on the part of the Company in the payment of the interest on these bonds, the trustee took possession of the railway for the purpose of foreclosing and selling the road. Under these circumstances, an arrangement was entered into between the preference bondholders, the ordinary bondholders, and a majority of the shareholders, as follows :—

(1) The present stock and all the bonds of the Company, except the preferential extension bonds, to be converted into new stock by the holders thereof at the following reduced rates :—(*a*) Bonds other than preferential extension bonds at 25c. in the dollar, with the exception of those now held by persons who are also at this date preferential bondholders, these latter to have the privilege of converting the ordinary bonds held by them at this date into new stock at 50c. in the dollar, but this privilege not to extend to bonds purchased by them subsequently to the passing of the Act of 1863. (*b*) The old paid-up stock to be converted into new stock at 10c. in the dollar. (*c*) The capital of the Company to be reduced to the amount of new stock required for such conversion, and in return for the privilege conceded to the preferential bondholders.

(3) The management of the road to be assumed by the preferential bondholders and their trustee as the Company, and the alleged rights of the preferential bondholders to foreclose and sell the road, to be waived and for ever extinguished without prejudice to their holding the first charge on the road, and on its revenues next after the municipalities, with all other legal remedies for the recovery of their interest and principal.

An Act was passed by the Legislature of Ontario, in 1867 and 1868, giving effect to this agreement. That Act specially provides that nothing in its terms shall in anywise affect the claims of the counties of Lanark and Renfrew, of the township of Elizabethtown, or the town of Brockville, upon the railway property.

The amount of paid up stock was $11,902.12, and a further sum of $165,-562.12, was turned over to the contractors, making the total paid-up capital stock $177,454.29. The amount expended on construction amount to 31st December, 1870, was $2,647,000. The gauge is 5 ft. 6 in.; weight of rail, (iron), 56 lbs. to the yard.

The revenue and expenditure, with per centage of same to gross receipts, were as follows for a series of years :—

	[illegible]	[illegible]	[illegible]	[illegible]	[illegible]
	$ c.	$ c.	$ c.	$ c.	$ c.
Revenue	[illegible] 10	[illegible] 94	57,778 [illegible]	[illegible]	[illegible] 12
Expenditure	34,[illegible] [illegible]	[illegible],771 [illegible]	[illegible] 71	[illegible]	[illegible]
Net revenue	19,378 [illegible]	14,[illegible] [illegible]	19,[illegible] 18	14,[illegible] 64	[illegible]
Per Cent. on gross receipts	64	[illegible]	[illegible]	75	[illegible]

The result of the operations for another series of years is shown thus :—

	[illegible]	[illegible]	1867.	[illegible]	[illegible]	[illegible]
Mileage of Trains	[illegible]	[illegible],725	[illegible]	No return.	1,[illegible]	....
" Locomotives	[illegible]	111,[illegible]	[illegible]	....	[illegible]	....
Passengers carried (number)	[illegible]	[illegible]	[illegible]	....	[illegible]	[illegible]
Freight, tons	[illegible]	[illegible]	[illegible]	....	[illegible]	[illegible]
Revenue.				....		
	$	$	$		$	$
Passenger Traffic	[illegible]	[illegible]	[illegible]		[illegible]	[illegible]
Freight "	[illegible]	[illegible]	[illegible]	....	[illegible]	[illegible]
Mails and Express	1,[illegible]	[illegible]	[illegible]	....	[illegible]	[illegible]
Sundries	1,[illegible]	7,[illegible]	[illegible]	....	[illegible]	[illegible]
Gross Earnings	[illegible]	[illegible]	[illegible]	....	[illegible]	[illegible]
Expenditure	[illegible]	[illegible]	[illegible]	....	[illegible]	[illegible]
Net Revenue	[illegible],915	[illegible]	[illegible]	....	[illegible]	[illegible]
Per ct. on gross receipts	[illegible]	[illegible]	[illegible]	....	[illegible]	[illegible]

The supply of rolling stock consisted in 1880 of 7 locomotives, 5 passenger cars, 138 freight cars, and 2 baggage cars.

DIRECTORS—The directors elected on Aug. 15, 1870, were: H. Abbott, Esq.; J. W. B. Rivers, Esq.; A. B. Dana, Esq.; B. Rosamond, Esq.; A. McArthur, Esq.; R. P. Cooke, Esq.

OFFICERS—H. Abbott, Esq., President; W. R. Worsley, Secretary and Treasurer; G. Lowe, jr., Accountant.

PRESIDENT AND MANAGING DIRECTOR:—H. Abbott, Esq., Brockville.

CHIEF OFFICE—Brockville, Ont.

ST. LAWRENCE & OTTAWA RAILWAY.

This road was projected by a party of Americans, mainly for the purpose of carrying lumber from the Chaudiere Falls to Prescott; was intended to be worked in connection with an American road then in course of construction, the northern terminus of which was Ogdensburg. The dictates of self-interest on the part of the then Directors is, no doubt, the reason for the important deviation of the original plan, by which the terminus was placed nearly three miles further down the river than when first contemplated. A consequence of this change is that the lumber traffic is not one-fourth what it would have been had the original project been adhered to. The name first given to this line was the Bytown and Prescott Railway; the first sod of which was turned in September, 1851. About £33,500 of stock was subscribed by the different municipalities interested, and over £20,000 by private parties. A reduction of some £10,000 had to be made from this amount on account of disputes and difficulties in making collections.

In March, 1853, the Company issued sterling bonds to the extent of £100,000, (payable in November, 1873, bearing interest at 6 per cent.,) which were sent to England to be negotiated. During the month of May of the same year, a contract was executed in Liverpool, England, with Ebbw Vale Iron Company, for 54,000 tons of iron rails, at £10.10s. per ton, payment to be made in the bonds of the Company at par. The equipment of the line consisted of 8 engines and 131 cars of all descriptions, which cost £45,000; £25,000 of that sum being payable in the Company's stock, and the remainder in money. The total cost of the road, 57 miles in length, and equipment, was over £250,000 sterling.

The Company received, under the provisions of the Grand Trunk Relief Act, £50,000 sterling.

During the years 1857 and 1858 the enterprise became very much involved, and various parties began to enforce their claims. The Ebbw Vale Iron Company seized their real estate, and the rolling stock was also taken possession of at the instance of other parties. The whole property was placed in the hands of a Receiver, appointed by the Court of Chancery. After a period of nearly four years, (January, 1862,) the matter was amicably settled, and the Receiver, by consent, removed. On the settlement, it was agreed that the Ebbw Vale Iron Company should be paid thirty per cent. of the gross earnings on account of their claim. This was paid from February until September, and amounted to $11,554.86. The decrease of traffic, owing to the stoppage of the work on the Parliament buildings, made it apparent that this large draft on the Company's revenue could not be much longer sustained; and the fact being so represented to the Ebbw Vale Iron Company, they allowed the payments to stand over, and the Receiver was re-appointed. An arrangement was made with the Grand Trunk Railway for the use of the track between the Prescott Junction and the St. Lawrence River, on condition that the Company should advance $7,000 to construct new works, and pay this Company at the rate of 25c. per ton, and make a fair allowance for passengers passing over this part of the line. An award was finally obtained from the Court of Chancery, in reference to the various claims upon the property; and, under the sanction of an Act of Parliament, the property was put up at auction, and sold to the holders of the first mortgage of £100,000; the price paid being represented by their claim, with interest and the cost of a seven year's law suit. An effect of this sale was to wipe out the second mortgage (to municipalities for $200,000); the third mortgage (given under provisions of Grand Trunk Relief Act, $242,383), and a large amount of floating indebtedness besides. The line runs from Prescott, on the St. Lawrence, to Ottawa, the capital of the Dominion; length of main line, 54 miles; sidings, 5 miles; total, 59 miles. Work was commenced in 1852, and completed in Dec., 1854; gauge, 4 feet 8½ inches; the bridges are of timber; that over the Rideau River has four spans of 100 feet each, and is supported on stone piers.

Rolling stock, 1869—Number of locomotives, 7; 12 first-class passenger cars, 5 second-class do, 53 box cars, 92 platform cars, 4 mail and express.

Mileage of passenger, baggage, box and platform cars ...	[illegible]
Number of passengers carried	[illegible]
Number of tons of freight carried	[illegible]

REVENUE, 1869.

Passenger traffic......	$53,064 82
Freight traffic..............................	56,790 00
Mails and sundries.........	8,770 35
Total income..	[illegible]

EXPENDITURE.

Operating..............	[illegible]
Interest, rent, &c.	19,077 [illegible]
	$107,087 71

DIRECTORS—(Elected May 9, 1870.)——William Quilter, London, England President; Thomas Reynolds, Ottawa, Ontario; Joseph Robinson, London, England; Thos. Robinson, London, England; Alexander Robert Eyre, London, England.

OFFICERS—Thomas Reynolds, Vice-President and Managing Director, Ottawa; R. Luttrill, Superintendent, Prescott, Ontario; C. Dame, Locomotive Superintendent, Prescott; F. A. Wise, Resident Engineer, Prescott.

WELLAND RAILWAY.

This line extends from Port Colborne, on Lake Erie, to Port Dalhousie, on Lake Ontario, a distance of 25 miles, and forms an important link in our great leading route of transportation from the upper lakes to the seaboard.

In 1859 the road was finally completed, and the total cost of the railway and equipment, up to last year, was $1,622,843. The line is laid with iron rails, 56 lbs. to the yard, and the gauge is the standard gauge of the Province —5 ft. 6 in. The bridges and works are first class.

Much the larger portion of the capital was raised in England, where it is still chiefly held. Quite recently a complete change took place in the *personnel* of the officers, owing to some difficulties which arose respecting the internal management of the Company's affairs. In consequence of this change, most of the documents relating to the earlier history of the Company are inaccessible at present, rendering the account we are able to give of the establishment and progress of this line necessarily very meagre.

The operations of 1870 are indicated as follows :—

Rolling Stock :—Number of locomotives employed, 5 ; 3 passenger cars, 147 freight cars, 2 baggage cars, 1 express car, and 8 gravel cars.

Total mileage of passenger trains in 1870, 31,300 ; number of passengers carried, 46,442.

Revenue:—Passenger traffic	$14,813 87
Freight traffic	50,626 58
Mails and sundries	8,293 63
Total	$73,734 08
Expenditure—gross	76,096 76

LONDON (ENG.) BOARD—Names of Directors elected in May, 1870 : J. W. Besanquet, London, England, Chairman of Board ; Major Kitson, R. B. Wade, Admiral Tyndale, Thomas Ogilvy, one vacancy in room of Thomas Brassey, lately deceased.

CANADIAN BOARD—Thomas R. Merritt, M. P., Chairman of Local Board, St. Catharines, Ont.; Hon. J. R. Benson, St. Catharines; John Brown, Thorold, Ont.

OFFICERS—Thomas R. Merritt, M. P., Chairman of Local Board of Management, St. Catharines, Ont.; Cornelius Stovin, Secretary, St. Catharines; William Pay, Superintendent, St. Catharines.

CHIEF OFFICE—St. Catharines, Ont.

NORTHERN COLONIZATIOTION RAILWAY.

(PROJECTED.)

A Bill was passed by the Quebec Legislature, and assented to on the 5th April, 1869, incorporating the Northern Colonization Railway Company. Power was also granted to extend from the main line, branch roads to St. Eustache, St. Scholastique, Lachute, Grenville; or to unite with the Carillon and Grenville railway, or to extend the line beyond the District of Terrebonne towards the city of Ottawa, so as to unite with such other railways as may hereafter be built by the "Canada Central Railway Company." Power is also given to extend branch roads as far north as St. Sauveur, St. Adele, St. Agathe, and easterly through the township of Kilkenny, to unite with the Lanoraie and Rawdon railway. The Company has also the special power of *buying*, transporting and *selling* firewood. The capital stock being $500,000, in fifty thousand shares of $10 each, with power to increase the capital stock to $3,000,000.

The Act authorized the construction of a line of railway from Montreal to St. Jerome, a flourishing village on the North River, and situated about 37 miles in a north-westerly direction from Montreal. The original intention seems to have been to construct a wooden railway for the chief purpose of supplying Montreal with cheap fuel. The project was encouraged by a promise of pecuniary aid from the Quebec Legislature to the extent of 8 per cent. on the cost of construction, such cost was not, however, to exceed $6,000 per mile. Three other projected wooden railways were included in the terms of this grant—the Quebec and Gosford, the Levis and Kennebec, and the Richmond, Drummond and Arthabaska Counties Railway.

The original scheme seems to have expanded very much, since it is now proposed to build a line along the north shore of the Ottawa River to Hull, opposite the capital—a distance of 120 miles—having the 4 ft. 8½ in. gauge, with rails weighing 60 lbs. to the yard, all important bridges being built on stone piers.

A land grant equal to 10,000 acres per mile was obtained at the last session of the Quebec Legislature, for the line to Aylmer, a point on the Ottawa River, about eight miles from Ottawa city, the aggregate grant being 1,350,000 acres. To this grant several conditions were appended :—(1) The land is not to be transferred to the Company till the road is completed. (2) A junction with the North Shore Railway must be effected at some point determined on. (3) That the lands might be granted as the work progressed, and (4) That one third of the directors should be appointed by the Government. For the purposes of this grant the road is divided into two sections, the one extending from Montreal to Grenville, and the other from Grenville to Aylmer. The grant to be made for the first named section is not to exceed 5,000 acres per mile until the whole line is completed, when the remaining 5,000 acres per mile will be added to that portion of the grant relating to the second section.

The cost for the 120 miles from Montreal to Hull is stated by Mr. Cyrus Legge, C. E., at $3,600,000, or $30,000 per mile. The Company's resources available for this distance (some of which are altogether prospective) are stated thus : 1,200,000 acres of land, estimated at $1 per acre ; bonus from the city of Montreal, $1,000,000 ; other municipal grants, $500,000 ; making a total of $2,700,000, and leaving $900,000 of the estimated cost to be raised by subscription or from other sources.

Directors—Hon. J. J. C. Abbott, Q. C., Hon. John Young, P. S. Murphy, Charles J. Coursol, Mayor of Montreal ; Duncan McDonald, Godfroi Laviolette.

Officers—Olivier Berthelet, President ; Louis Beaubien, M.P., Vice-President ; E. L. de Bellefeuille, Secretary.

Provisional Offices—15 St. Lambert Street, Montreal.

GLASGOW AND CAPE BRETON (N. S.) COAL AND RAILWAY COMPANY.

(PROJECTED.)

This Company was formed in London, England, to carry into effect an Act of the Legislature of the Province of Nova Scotia, passed on the 21st September, 1868, authorizing the construction of a railway from Sydney Harbour to Cow Bay, via Bridgeport, in the Island of Cape Breton, and for constructing warehouses, wharves, docks, and piers as may be necessary for the storing and shipment of coals and other articles. The capital is £100,000, in 10,000 shares of £10 each.

Mr. Featherstonhaugh, C. E., and an efficient staff, were sent out to inspect and report upon the proposed line, and to decide upon the necessary pier works at Sydney Harbour, whose estimates have been examined and approved by the Company's Engineer. The railway will be constructed on a gauge of 3 feet.

The length of the line, with its requisite sidings, will be about 21 miles, passing in its course through some of the most valuable coal-fields of the Cape Breton district, in which nine collieries are in full work, raising annually 575,000 tons of coal. At the present time most of these collieries can only export their coal in the summer, owing to the dangerous character of the coast to which alone they have now access.

The entire cost of construction and equipment of the line, including maintenance of way and works six months after the line is opened, as also the purchase of land, is estimated to amount to £82,000, and responsible contractors have guaranteed to complete the railway and pier by the end of October, 1871, within the price estimated by the Company's Engineer.

The Government of Nova Scotia have granted a lease for 75 years, as a bonus for the construction of the line, of one square mile specially reserved by them for this purpose, containing one of the most valuable seams of coal in the district.

The principal seam, known as the Phelan Vein, and varying from 7 to 8 feet in thickness, crops out upon the surface, whence it can be worked without the usual expense and delay of sinking shafts, the royalty being, until 1888, fivepence per ton, with the usual Government reservation for alteration after that period. The area is estimated to contain ten millions of tons of workable coal, and delivered at Sydney Harbour, is calculated to leave a profit of 3s. per ton. Simultaneously with the construction of the railway, the "reserve" coal-field will be opened up so as to be in a position to out-put 100,000 tons per annum so soon as the line is opened throughout; from the favourable position and easy dip of the seams this work can be done for £5,000.

Mr. Samuel, the Engineer, has furnished the following estimate of the traffic of this railway:—

200,000 tons of coal carried 13 miles, at 1½d. per ton per mile.........	£[illegible]
100,000 tons of coal carried at 1½d. per ton for 10 miles...............	6,[illegible]
Profit on sale of 100,000 tons from Reserve, 3s. per ton................	16,[illegible]
	£43,750

"As the line will have very favourable gradients, and coal for fuel is very cheap, and the rolling stock will be so constructed that the dead weight of the trains will be reduced to the minimum consistent with safety and durability, I believe that the traffic may be worked at an outside cost of 33 per cent., or one-third of the gross receipts of the carriage of coal, or ½d. per ton per mile, leaving a profit on the carriage of £19,167, and from sale of the reserve coal £16,000, making together a net revenue of £35,167, or 8½ per cent. on the proposed capital of the Company."

The following are the particulars of a contract entered into by the Company, which has been formed in England for the purchase of the charter powers conferred on a Nova Scotia Company :—

"The vendors have agreed to transfer to this Company the benefit of the Act of the Colonial Legislature, authorizing the construction of the railway and works with all its privileges, together with the lease of the coalfield, and to defray all the expenses of surveys, etc., etc., incurred up to the issuing of the prospectus, for the sum of £5,000 in cash and £5,000 in fully paid-up shares ; they are also to receive one-fourth of all profits after a dividend of ten per cent. per annum has been given to the shareholders. The agreement under which these benefits have been acquired is dated the 11th day of January, 1871, and made between Frederic Newton Gisborne of the one part, and Levi Elkin and Edward Ludwig Goetz, on behalf of this Company, of the other part."

DIRECTORS—Horatio L. Nicholls, Esq., Chairman, Southgate House, Southgate ; Thomas P. Baker, Esq., C. B. ; William Martineau, Esq., M. I. C. E. ; Herbert Heath, Esq., and Captain Powell, C. B., all of London, England.

ENGINEER—James Samuel, Esq., M. I. C. E., Westminster.

SOLICITORS—Messrs. Randall and Angier, 3 Gray's-in-place, Gray's-inn.

AUDITORS—Messrs. Ford and Smith, London.

SECRETARY—Mr. Walter Wright.

OFFICES—Great Winchester street, London, E. C.

TORONTO, GREY & BRUCE RAILWAY.

(UNDER CONSTRUCTION.)

Up to the time when this project was brought before the public, in 1867, the gauge of Canadian railways had uniformly been the standard or Provincial gauge of 5 ft. 6 in., except three lines—the St. Lawrence & Ottawa, the Montreal & Champlain, and the St. Lawrence & Industry, all of which are of the 4 ft 8½ in. gauge, being the same as that since adopted by the Great Western Railway. The idea of a railway with so narrow a guage as 3 ft. 6 in. was an entirely new idea with nearly everyone in this country, and like most other changes which conflict with interest and prejudice, excited a good deal of hostile criticism and not a little ridicule. Notwithstanding the fact that the application to the Ontario Legislature for a charter at the first session of that body in 1867-68, was supported by the names and influence of many of the leading merchants of Toronto, it was only carried through by a narrow majority and after a severe contest, first in the Railway Committee, and after-made on the floor of the House. The objection against the narrowness of the gauge has been urged with greater pertinacity, if not with equal ability, in the municipalities from which aid was being solicited. The disadvantage accruing to the promoters from this wide-spread objection was probably more than compensated by the consideration of cheapness in favour of a 3 ft. 6 in. line.

The agitation of the project—as well as also the sister enterprise, the Toronto & Nipissing Railway—had an important influence in re-directing public attention in this country to the advantages of railways, after the long period of repose in which railway progress was allowed to lie since the calamitous period of 1856-57. These schemes being regarded as practicable means of tapping two most important districts of Ontario, and placing them in close connection with the chief city of the Province, were eagerly seconded by the citizens of Toronto. The warmth of their support is best indicated by the grant of a quarter of a million of dollars as a gift to the Company, and by the subscription of three hundred and twenty thousand dollars of stock.

By the charter the Company is authorized to build a railway not less than 3 ft. 6 in. gauge (but of wider gauge if the directors at any time desire the change) from Toronto to Orangeville, thence to Mount Forest or Durham, thence to the border of the County of Bruce, and thence to Southampton, with a branch to Kincardine, on Lake Huron; also, a branch from Mount Forest or Durham or some point east thereof. The capital stock is $3,000,000, with power to increase the same in the manner provided by the General Railway Act, to be divided into 30,000 shares of $1,000 each. When $300,000 of the capital was subscribed, and ten per cent. paid, the Company could be organized. The management of the Company's affairs is in the hands of nine directors, each of whom must hold ten shares in the stock of the Company. Power is also given to issue bonds, the amount of which must not exceed the paid-up capital of the Company and the municipal bonuses actually expended in surveys or works of construction.

The clause relating to the carriage of cordwood reads thus: ("Clause 30.) The said railway Company shall at all times receive and carry cordwood, or any wood for fuel, at a rate not to exceed for dry wood 2½c. per mile per cord, from all stations exceeding fifty miles, and at a rate not exceeding 3c. per cord per mile from all stations under fifty miles, in full car loads; and for green wood at the rate of 2½c. per ton per mile. (Clause 31.) The Company shall further at all times furnish every necessary for the free and unrestrained traffic in cordwood to as large an extent as in the case of other freight carried over the said railway."

Owing to the refusal of the County of Grey to grant the aid asked for the construction of the proposed branch from Mount Forest to Owen Sound, that part of the scheme was changed, and the building of a branch from Orangeville direct to Owen Sound is now definitely decided upon, the necessary surveys being already in progress. The Company have agreed to complete the road to Owen Sound about November, 1872.

The immediate resources of the undertaking are:

BONUSES.

City of Toronto	$250,000
Township of Albion	40,000
Township of Caledon	45,000
Township of Mono	45,000
Village of Orangeville	15,000
Township of Amaranth	30,000
Township of Luther	20,000
Township of Arthur	35,000
Village of Mount Forest	20,000
County of Grey	300,000
Total of above bonuses	$795,000
Subscribed capital (50 per cent. paid up)	300,000
Bonds issued	160,000
Total	$1,250,000

The whole cost of the line, including rolling stock and equipment, is estimated at the low figure of $15,000 per mile—a sum which it is believed will not be exceeded.

The issue of bonds is limited by the charter to the amount of paid up stock and the bonuses actually expended in construction, but the directors do not anticipate a larger issue than at the rate of $6,000 per mile.

The following are the amounts received and expended by the Company, under the heads enumerated, up to 30th April, 1871 :

RECEIVED.

Calls from Stock	$131,400
Bonuses received from Trustees	237,634
Proceeds of Bonds issued	221,050
Bills payable	75,000

EXPENDED.

Preliminary expenses	$16,039
Right of way	40,437
Engineering	29,732
Stations	14,863
Construction	295,906
Iron and fastenings	231,500
Rolling Stock	107,062

The distances are as follows :

Length of line to Orangeville	50 miles
Orangeville to Owen Sound	70 miles
Orangeville to Mount Forest	39 miles
Total	159 miles

By an arrangement with the Grand Trunk Railway, this Company have permission to use the road bed of that railway for a distance of 9 miles from the city of Toronto, the amount of compensation to the Grand Trunk being a certain sum for each passenger and for each car load of freight traffic carried. By this means a considerable saving in cost of construction has been effected.

On the third October, 1869, the first sod was turned by Prince Arthur, and work was immediately thereafter commenced along the first section, to Arthur. A contract for the line from Orangeville to Mount Forest was awarded to Mr. Frank Shanly for earthwork, fencing, building all wooden bridges, furnishing and laying down ties, track-laying and ballasting. The contract for ballasting and track-laying from Weston to Orangeville was given to Messrs. Wardrop & Co.

By the 1st May, 1871, the track had been laid to Orangeville—49 miles—the grading and bridging were almost complete to Arthur village, a distance of 24 miles from Orangeville. There will be ten stations between Toronto and Orangeville.

The steepest grade going south is at the River Humber, where the ascent is at the rate of 66 feet per mile. Going north a steeper grade is encountered at the Caledon mountains, where the ascent is 106 feet to the mile. The sharpest curve is at the Humber, where the radius of curvature is but 492 feet. Upon the whole length of this line there are only three places where anything approaching to heavy works are met with—1st, at the crossing of the River Humber, in the township of Vaughan; 2nd, in the ascent of the Caledon mountain, extending over a distance of four miles; 3rd, at the crossing of the Grand River, in the Township of Amaranth. The only bridges of any size are those over the River Humber, consisting of six spans of 30 feet each, and one span of 38 ft. 6 in., built upon stone abutments and piers; the River Credit bridge in Caledon, one span of 46 feet and 12 trestle-work spans of 16 feet each; the Grand River bridge, two spans of 63 feet each, and five spans of trestle work, 25 feet each; and the Boyne Creek trestle bridge, one span of 40 feet and ten spans of 20 feet each. There are a few trestles, all but two of which are small in size, the exceptions being one of ten spans of 20 feet each over "Duncan's Ravine," and one of seven spans of 20 feet each over "Brown's Ravine."

The rails and rolling stock are all in proportion to the gauge of 3 ft. 6 in. The rails weigh 40 lbs. to the yard, and are of iron of English make. The locomotives range from 16 to 25 tons in weight, and were built in Bristol, England. A Fairlie engine of 42 tons is also in course of construction. The passenger cars are 35 feet in length, and weigh about 12,000 lbs., and will accommodate 40 passengers each. The platform cars are fitted with six wheels, and with radial axle-boxes, an arrangement by which the level of the floor is brought down to a distance of only two feet six inches from the rails. Box cars are also constructed, 15 feet in length, on four wheels, and are capable of carrying five or six tons each. The platform cars are 18 feet in length by 8 feet in width, and are capable of carrying a load of ten tons.

DIRECTORS (elected Sept., 1870)—John Gordon, Hon. John McMurrich, George Laidlaw, H. S. Howland, George Gooderham, John Shedden, Capt. Dick, R. Homer Dixon, Ald. Medcalf, John Norris.

OFFICERS—John Gordon, President ; Hon. John McMurrich, Vice-President ; W. Sutherland Taylor, Sec. and Treas. ; Edmund Wragge, Chief Engineer ; Allan Macdougall, Resident Engineer.

CHIEF OFFICE—Corner of Front and Bay Streets, Toronto.

TORONTO AND NIPISSING RAILWAY.

(UNDER CONSTRUCTION.)

The object of this undertaking is chiefly to establish direct communication between the city of Toronto and the extensive agricultural and lumbering region to the east of Lake Simcoe and the Georgian Bay. It has been warmly supported by the people of Toronto from its inception, for the reason chiefly that it must largely increase the trade of the district referred to with the city of Toronto. And, on the other hand, since it gives the inhabitants of the district a choice of markets it was warmly supported by them, and received their substantial aid in the shape of municipal bonuses.

The character of the proposed road is similar to the Toronto, Grey and Bruce. The charter was obtained at the same session of the Ontario Legislature—the session of 1867 and 1868. The amount of subscribed capital which must be obtained before the Company could organize was $150,000. In most every other respect the provisions of this charter are the same as those of the Toronto, Grey and Bruce, the cordwood clause being precisely similar.

The advantages of the light narrow gauge system, as adopted for this railway and the Toronto, Grey and Bruce, are stated thus :—

1st. The large comparative saving in first construction.

2nd. The large proportion of paying load to non-paying or tare weight of train.

3rd. The great reduction of wear and tear of permanent way, through advantage gained by light rolling stock

4th. Saving in reduced wear and tear of wheel tyres from reduced weight on each wheel.

5th. Large proportionate increased power of locomotives.

6th. Proportionate increased velocities gained by the light system.

7th. Greater economy in working traffic.

8th. Comparative increase in capabilities of traffic.

9th. Great advantages gained by the application of the Fairlie system of locomotive engines in concentrated power, equalization of adhesion of all the wheels to the rails, economy from reduced friction on wheel flanges, reduction of wear and tear to the permanent way, great saving in fuel, and economy in wages for given power secured.

Bonuses were given by the municipalities named as follows :—

City of Toronto	$150,000
Scarboro'	10,000
Markham	[illegible],000
Uxbridge	[illegible],000
Scott	15,000
Brock	[illegible],000
Eldon	44,000
Bexley	16,000
Somerville	15,000
Laxton, Digby and Longford	12,500
	$[illegible],500
Subscribed stock, 60 per cent. paid	[illegible],100
Debentures issued to 1st May, 1870	161,990
Total	$748,165

The following are the receipts to 1st May, 1871 :—

Cash from calls on Stock	$ [illegible],920 00
Cash from Trustees on account of Bonuses	[illegible],424 [illegible]
Cash for Bonds issued	138,[illegible] 00
Notes payable	[illegible],[illegible] 36

Amounts disbursed on the following accounts :—

Construction	$375,[illegible] 55
Engineering and Surveying	36,610 [illegible]
Rolling Stock	[illegible],277 48
Right of Way	27,[illegible] [illegible]
Preliminary Expenses	16,[illegible] [illegible]

It is fortunate that the route of the railway runs through a most favorable country. There are really no heavy works on the line ; the rolling character of the country in the township of Uxbridge necessitates a good deal of excavation. The average number of yards of earthwork is 9,000 yards per mile. The only bridge of any size between Toronto and Uxbridge is that over the River Rouge, near Unionville, in the township of Markham, and which consists of three spans of 44 feet each, and four spans of 16 feet each, the whole structure is founded upon rock elm piles. The bridge over the north-west bay of Balsam Lake, near Coboconk, is the largest structure on the road ; it has three spans of 106 feet each, and 5 of 32 feet, being a total length of 478 feet. The other bridges which are already executed are, three small bridges in the township of Scarboro', all over the Highland Creek or its branches, and two more over feeders of the River Rouge, in the township of Markham. There will be three small bridges in the township of Brock, over the Beaver Creek ; and, with the exception of a trestle bridge at Markham seven spans of 20 feet each and a few short trestles of three spans of 16 feet each, here and there, this constitutes the whole of the bridge-work.

Shortly after ground was broken, a contract was let to Messrs. John Ginty

& Co. for the earthwork, and over the entire section of 86 miles to Coboconk, conditional on the granting of the expected bonuses from the townships along the line from Uxbridge to Coboconk. The contract for fencing and ties for the 32 miles to Uxbridge village was given to Mr. Edward Wheler. Messrs. Ginty & Co. relinquished their contract for the portion of the line north of Uxbridge, and it was given by the Company to Mr. Duncan McRea, M.P.P., of Eldon, on the same terms as were made with Messrs. Ginty & Co. Messrs. Wardrobe & Co. have the contract for track-laying to Uxbridge, a distance of 32 miles. E. Wheler has the contract for sections, tanks and engine-sheds from Uxbridge to Coboconk.

The line is finished to Uxbridge, and will be formally opened for traffic on the 1st of July next. The remainder of the line is more than half completed, and will be ready for opening, it is hoped, in the latter part of this year.

The steepest grade going north is one foot in fifty; going south, one foot in sixty. The sharpest curve is at Uxbridge, and has 800 feet radius. The passenger cars are 35 feet in length, and capable of holding 40 passengers each. The platform cars are thirty feet in length by eight feet in width, and are capable of carrying ten tons each. The box cars are 15 feet in length by eight feet in width, and will carry from five to six tons. About 180 or 190 cars are being turned out by a Toronto firm, Messrs. William Hamilton & Son. The locomotives are made by the Canadian Engine and Machinery Company, Company, at Kingston, a Fairlie engine of 42 tons weight and another large freight engine are being made in England.

The gauge being 3 feet 6 inches, the rails are correspondingly light, being 40 lbs. to the yard. The iron was purchased in England, with a guarantee for seven years, at the rate of £8 5s. per ton.

Directors—(Elected Sept., 1870)—John Shedden, Wm. Gooderham, jr.; J. C. Fitch, Joseph Gould, T. C. Chisholm, George Laidlaw, James E. Ellis, Hugh Macdonald, John Gardner and William Adamson.

Officers—John Shedden, President; William Gooderham, jr., Vice-President; James Graham, Secretary and Treasurer; Edmund Wragge, Chief Engineer; J. C. Bailey, Resident Engineer.

Chief Office—Corner of Front and Bay streets, Toronto, Ontario.

WELLINGTON, GREY AND BRUCE RAILWAY.

(UNDER CONSTRUCTION.)

The charter authorizes the building of a line of railway from Guelph to some point in the county of Bruce, with the object of, at some time, extending to Lake Huron.

The capital stock of the Company is $[illegible], and bonuses have been granted as follows:—

City of Hamilton	$ [illegible]
Village of Elora	10,000
Village of Fergus	15,000
Nichol	10,000
Township of Peel	40,000
Township of Maryborough	40,000
Township of Wallace	25,000
Township of Minto	70,000
Township of Howick	20,000
County of Bruce	250,000
Total	$461,000
Add Capital Stock	30,000
	$491,000

This project received the most active and energetic support of the citizens of Hamilton, to whose efforts the progress made (together with the support of the Great Western Railway Company) is chiefly due. The chief object of the line is to divert the trade of Wellington and Bruce Counties to Hamilton, as far as that is possible. The line was opened from Guelph to Elora on the [illegible] September, 1870, a distance of 16 miles; and from Elora to Alma, in December, 1870, a further distance of 5 miles, making in all 21 miles. The distance to Southampton on Lake Huron is 98 miles from Guelph. A contract has been let to Mr. Hendrie, of Hamilton, for the extension of the line to the county of Bruce. This Railway must be completed to Southampton, in the County of Bruce, in June, 1872, or the large bonus granted by this county will be forfeited and lost to the Company. By an agreement made with the Great Western Railway, dated June 15th, 1869, that Company undertakes to stock the line and work it as soon as completed to the satisfaction of the Manager of the Great Western; and to regulate the rates of freight and all other charges; the lease to continue for one thousand years; the Great Western Company to pay over to the Wellington, Grey and Bruce Company 30 per cent. of the gross traffic; an account to be kept of the traffic interchanged between the two lines, of which a sum amounting to [illegible] per cent. of such receipts shall be appropriated annually to the purchase of the bonds of the Wellington, Grey and Bruce Company, but from these receipts are to be deducted every year the sum of $76,061, which is the average traffic in and out of Guelph for the past three years, the intention being to pay the foregoing 20 per cent. only on the increase of traffic derived from the Wellington, Grey and Bruce, the latter Company being bound to issue not more than $[illegible],000 of bonds for every mile of railway constructed. By a subsequent agreement of the 3rd of June, 1870, the Great Western Company undertake to purchase bonds of the Wellington, Grey and Bruce Company to the amount of $12,000 per mile of railway constructed, and to that increased amount the issue of bonds by the Wellington, Grey and Bruce Company is now limited.

The line is being constructed under the supervision of Mr. George Lowe Reid, the Engineer of the Great Western, and when completed will be a most valuable branch of the Great Western.

OFFICERS—Col. McGiverin, President, Hamilton; George Lowe Reid, Esq., Chief Engineer, Hamilton; James Osborne, Secretary, Hamilton.

CHIEF OFFICE—Hamilton, Ontario.

KINGSTON AND PEMBROKE RAILWAY.

(PROJECTED.)

This line is to run from Kingston, on the St. Lawrence, to the town of Pembroke, on the Ottawa River. The City Council of Kingston have introduced a by-law, granting a bonus of $300,000 in aid of the enterprise. The promoters expect to receive in the way of bonuses, besides $300,000 from the city of Kingston, $150,000 from the county of Frontenac; county of Renfrew, $400,000; town of Pembroke, $50,000, and from the Ontario Government, under the Act to aid railways running toward our inland waters, about $600,000, making a total of a million and a half of dollars. The length of the road will be about 140 miles, and the total cost about $2,000,000.

CHIEF OFFICE—Kingston, Ontario.

CANADIAN PACIFIC RAILWAY.

(PROJECTED.)

The people of the Dominion are pledged to the construction of this railway, by the terms of the arrangement made with British Columbia for the admission of that Province into the Dominion. The resolution relating to the railway is as follows:—

"The Government of the Dominion undertake to secure the commencement simultaneously, within two years from the date of union, of the construction of a railway from the Pacific towards the Rocky Mountains, and from such point as may be selected east of the Rocky Mountains towards the Pacific, to connect the seaboard of British Columbia with the railway system of Canada; and further, to secure the completion of such railway within ten years from the date of union. And the Government of British Columbia agree to convey to the Dominion Government, in trust, to be appropriated in such manner as the Dominion Government may deem advisable, in furtherance of the construction of the said railway, a similar extent of public lands along the line of the railway throughout its entire length in British Columbia, not to exceed, however, 20 miles on each side of the said line, as may be appropriated for the same purpose by the Dominion Government from the public lands in the Northwest Territories and Province of Manitoba; provided that the quantity of land which may be held under the pre-emption right or by the Crown grant within the limits of the tract of land in British Columbia to be so conveyed to the Dominion Government shall be made good to the Dominion from the contiguous public lands; and provided further that until the commencement, within two years, as aforesaid, from the date of the union, of the construction of the said railway, the Government of British Columbia shall not sell or alienate any further portions of the public lands of British Columbia in any other way than under the right of pre-emption requiring the actual residence of the pre-emptor on the land claimed by him. In consideration of the land to be so conveyed in aid of the construction of the said railway, the Dominion Government agree to pay to British Columbia, from date of union, the sum of $100,000 per annum, in half-yearly payments in advance."

Sir George Cartier, as Leader of the Government, explained the views of Ministers in a speech delivered before the House of Commons at the session of 1871. He said there was a good deal of misapprehension with regard to what the Government intended to do respecting this railway. The Government did not intend to build the railway themselves, but by means of companies that would have to be assisted principally by grants of one dollar lands. The land which British Columbia would contribute for this purpose was valued at one dollar an acre, which would amount to $15,360,000. For this the Government would undertake to pay $100,000 a year to British Columbia, which was the interest at five per cent. on two million dollars. That was to say that, in the purchase of these two millions of acres, Government would be a gainer to the extent of $13,360,000 with which to assist the railway that would be undertaken. The Government insisted upon that as a sine qua non condition. The land must be under their control in order to aid the railway. It was estimated that the length of the road to be built from Lake Nipissing to Victoria was about 2,500 miles; twenty miles on each side of the road would give 64 millions of acres to be used in aid of the line. About 600 or 700 miles of the line would be within the Province of Ontario; and he had reason to believe the Government of that Province would have the liberality to give them, not twenty miles on each side, but at least every alternate block on each side. That would be a contribution of about nine millions of acres. Lake Nipissing would be a junction where the lines both for Ottawa and Toronto could meet. The contribution of land itself would be almost enough to build the railway. If any money subsidy was to be given, the Government would never go so far in that direction as to necessitate any increase of taxation in this country. The Government stated its determination that for the building of this railway land grants would principally be relied on. If any subsidy would be given, it would be a moderate one, and one that would not require any further taxation on the tax-payers of the Dominion.

The terms of the resolution given above would require positively the completion of the line within ten years from the date of the union. As explained by the Government, it is apparent that the intention is to construct the railway within that time, unless the undertaking should have to unduly press upon the finances of the Dominion. The mode of construction is expressly limited by the subjoined resolution, passed by the House of Commons just before rising :—

"Resolved, that the railway referred to in the address to Her Majesty concerning the Union of British Columbia with Canada, adopted by this House on Saturday last, April instant, should be constructed and worked by private enterprise, and not by the Dominion Government ; and that the public aid to be given to secure that undertaking, should consist of such subsidy in money, or other aid, not unduly pressing on the industry and resources of the Dominion, as the Parliament of Canada shall hereafter determine."

The proposed gauge is 4 feet 8½ inches.

THE RICHMOND, MELBOURNE & MISSISQUOI VALLEY RAILWAY.

(PROJECTED.)

This road would extend from Richmond Junction, Eastern Townships, via Lower and Upper Melbourne Villages, through nearly the centre of the Township of Melbourne, near Kingsbury, through Brompton Gore, Ely via Lawrenceville, through North and South Stukely, Bolton and Potton, to a point a little south of Masonville, where it would meet the Missisquoi River Road, which is soon to be completed from St. Albans to Newport, Vermont. The South Eastern Counties' Road, now being built from West Farnham, would also connect with it, within or near the Township of Potton. This proposed route from Richmond, less than 50 miles in length, would pass through the very heart of the Eastern Townships.

THE MASSAWIPPI VALLEY RAILWAY.

This railway has been leased to the Passumpsic Company for 999 years. This line is to connect the Connecticut and Passumpsic River Railway with the Grand Trunk at Lennoxville, and is about 34 miles in length, including the spur of 3½ miles, running to Rock Island, Stanstead. The line was opened on 1st July last. $166,000 of stock was subscribed in Canada and paid in gold and an equal amount furnished by the Passumpsic Railway Company, making $330,000 cash stock. The contractors are to take as part payment $70,000 of stock, $400,000 of bonds to be issued by the Massawippi Company, which the Passumpsic Company endorse and guarantee and provide for. The road and real estate from the line to Lennoxville is mortgaged for security of these bonds, and to aid in the redemption, a like amount of stock is issued. The Passumpsic Company undertakes to build, equip and run the Massawippi Valley road, and to lease the same, paying interest on the bonds, $24,000 in gold, to the holders, in semi-annual payments. The Passumpsic Company also undertake to pay to the stockholders in the Massawippi Railroad Company, from the earnings of both roads, equal dividends per share with that paid to the stockholders in the Passumpsic Railway Company. The total of the dividends appropriated to the Massawippi Railway Company stockholders not to be less than one-fifth of the whole sum divided to both Corporations. The gold value of the Passumpsic Railway is estimated at and put into the partnership thus in effect formed at $3,200,000, and the Massawippi Valley Railway is put at $800,000. Both roads will be operated by the Passumpsic corporation, in connection with the Massawippi corporation. The spur to Rock Island is built and worked in the same way as the main line. The contractors received $330,000 cash and $70,000 in stock and proceeds of the road, and $400,000 in bonds. The $165,000 contributed on this side is composed of subscriptions in Stanstead and vicinity, $100,000; in Hatley debentures, $15,000; and in Ascott debentures, $40,000, with some subscriptions in the vicinity to pay for the right of way over and above what the $15,000 in stock would meet, and for the preliminary expenses.

It has very recently been decided to run this line into Sherbrooke, E. T. by laying a third rail on the Grand Trunk from Lennoxville; stations will also be built there sufficiently commodious for the traffic of both lines.

The gauge is 4 feet 8½ inches.

Chief Office and Address—Lyndonville, Vt.

QUEBEC AND GOSFORD RAILWAY.

This is a line of 27 miles in length, from the city of Quebec to the village of Gosford, and is the only wooden railway in the Dominion at present. It was constructed by Mr. Hulbert, who has had experience as a contractor and operator of wooden railways in the United States. He commenced work on the line in September, 1869, and completed it in December, 1870, or a fortnight before the time required by his contract.

The road has quite a substantial appearance. The ties are heavy, and are laid so as to measure 2 ft. 4 in. from centre to centre. The rails are strips of seasoned maple, 14 feet long, 7 inches by 4 inches, notched into the seven sleepers over which each rail extends, and wedged hard and fast without the use of nails or iron fastenings of any kind. The 4 ft. 8½ in. gauge was adopted for this line. The rails are expected to last five or six years, and are perfectly safe so long as the wood remains sound. As steep a grade as 250 feet to the mile is encountered at one point on the road—an impossible grade for an iron railway. The bridge across the Jacques Cartier River, and the northern approach to it, is the most expensive work on the road, having cost $12,000. It crosses the river immediately above a beautiful fall of 30 feet. The river is 200 feet wide, and is crossed with two spans. The top of the bridge is 66 feet above the water. The trestle work on the northern end of the bridge is 1,250 feet long, and carries the road over the tops of the large forest trees growing in the valley beneath.

The cost of the road, including right of way and rolling stock, has been $6,000 per mile.

The stockholders are entitled to one cord of hardwood at cost price for every $10 paid in.

The rolling stock consists of one locomotive, twenty-five platform cars, four second-class passenger cars, and one box freight car. The wheels of both locomotive and cars have a greater diameter than those on iron roads, and are also broader, so as to cover the whole thickness of the wooden rail. The locomotive was built at the Rhode Island Engine Works ; it arrived in Quebec on the 23rd of June, and commenced running about the middle of July, and continued until late in December ; during the whole time it worked most satisfactorily. It was running an average of 100 miles per diem for 140 days, making a total distance of about 14,000 miles. The cost of the locomotive, including transport, haulage, &c., is $3,396.47. It weighs 21 tons, wooded, without the tender, and 28 tons with the tender. It will draw 75 per cent of what the same power could do on an iron road.

In March, 1870, a contract was let to Mr. S. Peters, of Quebec, for the construction of 30 platform cars, at a cost of $310 each; the cars were to have been handed over by the 15th July. They were not, however, delivered until late in the season, owing to the destruction of Mr. [illegible]'s foundry by fire, where the wheels were being made. The first pattern wheels having been found to be too slight, wheels of a heavier pattern have been substituted at an extra cost. One of the cars has been converted into a box freight car, at a cost of $190, and four others into temporary passenger cars, at a cost of $140 each. The total cost of these cars, including $147 for haulage, amounts to $10,464.85.

The total amount spent by the Company for all the purposes of its incorporation amounts to $140,864.60, up to 7th Feb. 1871. That amount was received from the following sources:

Paid by shareholders on their shares....................	$[illegible],4[illegible] [illegible]
Paid-up stock issued to contractor, as per contract.......	20,110 00
Directors' personal notes, redeemed with Govt. subsidy...	47,495 00
Company's note..	2,719 00
Interest...	464 46
Total..	$1[illegible],164 85

The Company's liabilities amounted at that time to $14,000 00, besides some unsettled claims for right of way. The Quebec Government has paid the sum of $48,171 20, for the subsidy due the Company in virtue of the 32 Vic., cap. 52, in cash, in full of the whole amount of the subsidy, instead of debentures or by twenty yearly payments.

Power was obtained at the last session of the Quebec Legislature to extend the line as far as Lake St John, and the Government have explored and have undertaken the work of locating that extension.

Respecting the solidity of the road, that would be quite satisfactorily settled by the tests which have been applied. At least 1,000 trains have run over the road without perceptible detriment to the rails or any of the bridges or other works.

The Company seem to have met with a good deal of difficulty in the collection of instalments due by the shareholders. Out of 1,241 shareholders, 780, representing 1,614 shares, had paid nothing in February last. A good many defaulters were sued, in most cases with success. The Board recommended that those 1,614 shares should be declared forfeited. Under these circumstances of difficulty, an offer was made by Mr. Hulbert to complete the road by an expenditure of $24,000 upon it, and work it for a term of years, paying the shareholders six per cent. interest on their capital. These favourable terms—for the Company, at least—were accepted, and the road is now being operated under an agreement arranged on that basis. The additional expenditure is mostly for fencing, stations, workshops and engine-houses, and additional rolling stock.

President—H. G. Joly, Esq.

Engineer—Mr. Rickon.

Chief Office—Quebec, P. Q.

SAULT STE. MARIE RAILWAY.

(PROJECTED.)

A charter was obtained at the last session of the Dominion Parliament, granting the necessary powers to build a line of railway from the village of Sault Ste. Marie, in the district of Algoma, to connect with the projected railway in the Province of Ontario, at or near Lake Nipissing, and to extend a branch therefrom to connect with the Toronto, Simcoe and Muskoka Junction Railway at or near Bracebridge, in the County of Victoria. Power is also granted to bridge the River St. Mary, and there effect a junction with lines in the United States. The corporators are—J. S. McMurray, F. W. Cumberland, J. B. Robinson, S. B. Harman, Angus Morrison, W. M. Simpson, Anson G. P. Dodge, Eli C. Clarke, S. E. Marvin, John McIntyre, John M. Hamilton, James Bennett, Walter McCrae, T. W. Herrick and J. J. Vickers, and these gentlemen are, by the Act, made the first Directors of the Company. The capital is fixed at $10,000,000. When $10,000,000 are subscribed, and 10 per cent. paid up, new Directors may be elected by the shareholders.

One of the chief objects of the promoters is to establish a connection between the railway system of Canada and the Northern Pacific, now under construction. It is believed that this connection will bring a large amount of through trade over the Toronto and Muskoka and the Northern to Toronto, where it can either be moved to New York by the Great Western, or to Montreal by the Grand Trunk. It would, besides, give a winter and summer route—all rail, when the branch of the Northern Pacific is completed to Pembina, as it soon will be—to the Red River Territory. In this way it would serve as a temporary substitute for a Canadian Pacific Line proper for the distance between Toronto and Manitoba. It would also afford an outlet for the product of the extensive mills along the north shore of the Georgian Bay, which are now entirely shut in during the winter. The promoters think that so important a link in our railway system should receive the maximum rate of subsidy from the Ontario Government provided by the Act of last session—$4,000 per mile—and a liberal land grant beside.

The distance from St. Marie to Bracebridge is 280 miles. The road, if built, will be of the 4 feet 8½ inch gauge.

MONTREAL AND CHAMPLAIN RAILWAY.

On the 25th of Feb., 1832, the Champlain and St. Lawrence Railway obtained their charter. The capital of the Company was £50,000, in shares of £50 each, with power to increase the shares to £65. The charter underwent several successive amendments. The road was constructed with wooden rails and thin flat bars of iron spiked upon them. It was in the first instance built from St. Johns to Laprairie; this section was opened in July, 1836. Subsequently, in order to give a closer connection with the City of Montreal, the northern terminus was transferred from Laprairie to St. Lambert, immediately opposite Montreal. This change was accomplished in January, 1852. In August of the previous year, the line had been extended from St. Johns to Rouse's Point, making a total distance from Montreal of 49 miles. The length of sidings, &c., is 5.65 miles, which gives a total mileage of track 54.65 miles.

This road is now leased and operated by the Grand Trunk Railway Company. The net revenue due the Montreal and Champlain Company under the agreement for the year ending 31st Dec., 1869, amounted to $105,856, being an increase on the former year of more than 16 per cent., which was more than sufficient to pay the interest on the Company's bonds, and the dividend on the preferred stock, interest on the sinking fund and incidental expenses, besides reducing a debit against the revenue account over $10,000. The amount standing at the credit of the sinking fund is $30,364.

CAPITAL ACCOUNT, 31ST DEC., 1869.

Consolidated Stock	$1,130,276
Preferred Stock	404,600
First Mortgage Bonds	80,300
Consolidated Loan	802,513

PER CONTRA.

Railway Property	$2,254,376
Fuel and Stores Stock	33,141

PRESIDENT—Hon. James Ferrier.

CHIEF OFFICE—Point St. Charles, Prov. of Quebec.

HARRISBURG & BRANTFORD RAILWAY.

(PROJECTED.)

This is a short line of seven miles, projected from Harrisburg to Brantford. The road is to cost $150,000, and will be built by the Great Western Company. A contract has been let to Mr. Hendrie, of Hamilton, and the work of construction will doubtless be proceeded with as early as practicable.

CARILLON & GRENVILLE RAILWAY.

This Company obtained their charter on the 24th June, 1848. Their capital is £60,000, in shares of £25 each. The line was run from Carillon to Grenville, a distance of $12\frac{3}{4}$ miles, and was opened for traffic during the month of October, 1854. This road is operated during the summer months only. The cost of road and equipment is $110,000. The rolling stock consists of 2 locomotive engines, 6 passenger and baggage cars, 2 box and 4 platform cars.

CHIEF OFFICE AND ADDRESS—Grenville Post Office.

ST. LAWRENCE & INDUSTRY RAILWAY.

On the 28th July, 1847, a charter was granted to a Company formed to construct a railway from Lanoraie, district of Montreal, to Industry village, a distance of 12 miles. This road was completed and opened for traffic in the month of May, 1850, and is operated during the summer months only. The cost of construction and rolling stock amounts to about $56,000. The rolling stock consists of 3 locomotive engines, 2 passenger and baggage cars, and 9 other cars.

CHIEF OFFICE AND ADDRESS—Industry, Province of Quebec.

MIDLAND RAILWAY OF CANADA.

(FORMERLY PORT HOPE, LINDSAY AND BEAVERTON.)

This Company was originally chartered on the 29th December, 1846. On the 14th December, 1858, power was given to build a branch through the townships of Cavan, Emily, Manvers, Ops and Mariposa, and thence to some convenient point on the line of the Ontario, Simcoe and Huron Railway—powers which were never fully exercised.

The Company were aided with large municipal subscriptions.

The Town of Port Hope gave, in all	$[illegible],000
Township of Hope	60,000
Township of Ops	80,000
Town of Peterborough	100,000
Total up to 1864	$980,000

Additional sums were subsequently granted.

On the 13th March, 1857, a lease of the road was given to Messrs. Tate & Fowler, one of the conditions of the lease being that this firm were to build the Peterborough branch. In aid of this work the town of Peterborough granted £30,000 and Port Hope £10,000, these towns taking a mortgage on the lease as security. By this mortgage the lessees were to pay Peterborough annually the sum of £1,800, and Port Hope, £600. By an Act of 18th Oct., 1860, the amount secured to Peterborough was reduced to £19,700.

By an Act of the 30th June, 1864, the purchase of the Port Hope Harbour was authorised, the bonds and debentures of the Harbour Company being exchanged for bonds of the Railway Company, bearing various rates of interest.

In 1866 the Company were authorized to purchase the Millbrook branch, and to issue preference bonds therefor to an amount not exceeding £110,000 stg. It was provided that these bonds should not be issued without the consent in writing of the persons with whom were deposited the then existing mortgage bonds of the Company as collateral security for the due payment of certain bonds given by Henry Covert, Esq., (now President of the road) for the purchase made by him of the then existing bonds of the Company. By the same Act the Company were authorised to issue preference bonds to the township of Hope in exchange for the stock held by the municipality sufficient to secure the annual sum of $1,542; to the township of Ops, in the same manner, bonds to secure $296.75; and to the town of Lindsay for $296.75 annually. All these sums are payable on the 1st December in each year, and constitute a first charge on the railway. It was also provided by the same Act that any stockholder could transfer his stock to the Company and receive in exchange therefor first preference bonds to the amount of 50 per cent. of such stock.

On the 23rd January, 1868, an arrangement between the town of Port Hope and the railway Company was legalized, by which the town was authorized to transfer sterling debentures of the Port Hope Harbour Company to the railway Company, the object being to aid the extension of the railway from Lindsay to Beaverton, by granting the Company the sum of £30,000. During the work of construction the railway handed over to a private individual $30,000 of mortgage bonds as security for the completion of the road to Beaverton on or before September, 1871. The Line was formally opened to Beaverton in January, 1871.

By an Act of 24th Dec., 1869, the name of the Company was changed to "The Midland Railway of Canada." Authority was also given to build a branch line from some point in the township of Mara through the township of Rama to the river Severn. Power was also granted to issue £100,000 of bonds. The Township of Thorah had loaned the Company the sum of $50,000 to extend the line to Beaverton, and by this Act the Company were authorized to give that township a lien on the railway in perpetuity for the sum of $1,500 per annum, being interest at the rate of 3 per cent. on the amount loaned, and payable on the 15th of June in each year.

An Act of the last session of the Ontario Legislature recited that there were at that time (Feb., 1871) outstanding first preference bonds to the amount of £110,000 stg.; second preferences to the amount of £125,000 stg., and £100,000 stg. of bonds then authorized but not issued, and gave power to substitute for these issues new consolidated six per cent. bonds to an amount not exceeding £335,000 stg., these bonds to form a first charge on the line.

The gross earnings in 1867 were $234,476.98; 1868, $232,904.10; 1869, $225,851.23; 1870, $242,157.22. The working expenses, including maintenance of way, in 1870, are stated at $128,930.03, or 53.24 per cent. of the gross earnings, leaving net revenue, $113,227.19. During the four years ending Dec., 1870, the gross sum of 308,000 was expended in improving the line. The operating expenses for the years named were: 1867, $100,000; 1868, $107,000; 1869, about $109,000; 1870, $128,930, which, added to the expenditure for the improvement of the line during these four years ($308,000), as given above, gives a total outlay in the four years ending Dec., 1870, of $753,000 in round numbers.

STATEMENT SHOWING TONNAGE OF THE PRINCIPAL ARTICLES OF FREIGHT.

	1867.	1868.	1869.	1870.
Square Timber, cubic feet	66,378	75,833	11,278	788,640
Sawed Lumber, feet B. M. ...	71,892,950	72,502,050	64,043,450	71,225,600
Wheat, Bushels................	246,277	200,649	262,626	249,752
Other Grains, bushels..........	173,427	128,407	131,447	151,914
Flour and Oatmeal, bushels...	43,312	36,907	44,567	26,334
Potash, barrels	237	164	184	158
Pork, barrels	1,510	2,042	3,889	1,213
Other Freight, tons............	16,966	17,681	15,296	19,540
Total No. of tons carried......	197,324	190,005	176,448	195,698

STATEMENT OF PASSENGER TRAFFIC.

	1867.	1868.	1869.	1870.
No. of Through Passengers...	32,064	36,960	30,470	38,947
" Way " ...	13,963	17,210	16,645	14,271
Total number of Passengers...	46,027	54,170	46,890	52,670
Miles Travelled, Thro' Pass.	970,160	272,460	1,168,644	1,212,840
" " Way "	112,812	1,072,040	307,977	210,907
Average by each Pass., miles	23[illegible]	24[illegible]	26[illegible]	27[illegible]

The following is a statement of the Earnings and Expenditures for the year ending 31st December, 1870 :—

EARNINGS.

Freight	$196,096 23
Passengers	43,210 61
Mails	2,848 88
Total	$242,157 32

EXPENDITURES.

Operating Expenses—

General management	$4,736 04	
Interest, Agency and Travel	12,910 31	
Traffic Depar't, including Station Agents	9,802 66	
Train and Water Service	15,616 85	
Office Expenses, Printing, Taxes, &c.	3,130 79	
Fuel, Oil and Waste	17,396 66	
Rolling Stock	20,940 28	
Machinery and Tools	2,746 55	
Track	24,386 02	
Bridges and Culverts	1,497 06	
Buildings and Fences	2,036 80	
Miscellaneous	5,101 02	
		126,930 03
Nett Revenue		$115,227 19

The line runs from Port Hope, on Lake Ontario, to Beaverton, on Lake Simcoe, a distance of 66 miles, with a branch from Millbrook to Peterborough, of 13 miles, making the total length of line opened, 79 miles.

ROLLING STOCK (Jan., 1870)—11 locomotive engines, 5 passenger cars, 6 mail and express cars, 192 freight cars, 40 box cars, 2 stock cars, 150 platform cars, also 3 service cars.

DIRECTORS—Henry Covert, Port Hope ; Sidney Smith, Peterboro' ; Lewis Moffatt, Toronto ; William Cluxton, Peterboro' ; D. E. Boulton, Cobourg.

OFFICERS—Henry Covert, President; William Cluxton, Vice-President; Joseph Gray, Sec. and Treas.; —— Taylor, Superintendent; G. A. Stewart, Chief Engineer; G. L. Fisk, Road Master.

CHIEF OFFICE AND ADDRESS—Port Hope.

CANADA CENTRAL RAILWAY.

This Company was chartered by Act of Parliament of Canada, assented to 18th May, 1861. The Act was an amendment of a previous Act "To encourage the construction of a railway from Lake Huron to Quebec." The Company obtained power to construct a line of road from Lake Huron to the City of Ottawa, via Pembroke and Arnprior, and from Ottawa to Montreal. The North Shore, the Carillon and Grenville and Canada Central Railway Companies may amalgamate. These Companies may also share in the grant of land given fo the above object in the manner prescribed by the Act. The authorized capital is $7,000,000, in 70,000 shares of $100 each. Power is given to issue bonds to the amount of one-half the capital. As soon as the railway is completed 20 miles, the Company may have a share in the land grant. On the 15th of August, 1866, the charter was amended, and a divergence in the line authorized between Ottawa and Pembroke, which permitted the Company to build their road at a distance from the Ottawa River not greater than 25 miles.

The line was built by Mr. H. Abbott, of Brockville, to Carleton Place, a distance of 28 miles from Ottawa, and was formally opened for traffic on the 15th September, 1870. It passes through a good country, now devastated, however, by the late disastrous fires. It connects with the Brockville and Ottawa line, running to Brockville and to Sand Point, on the Ottawa River, and is operated in connection with the Brockville & Ottawa Railway. The traffic earnings for the three months ending December 30th, 1870, amounted to $7,554.

DIRECTORS (Jan., 1870)—J. G. Richardson, President; S. Abbott, Vice-President; R. W. Scott, Hon. John Hamilton, Messrs. Askworth, Lowe and Rivers.

GENERAL MANAGER—H. Abbott, Esq., Brockville, Ont.

CHIEF OFFICE—Brockville, Ont.

PROVINCE LINE RAILWAY.

On the 24th of June, 1846, a charter was granted to the Lake St. Louis and Province Line Railway Company, with a capital of £160,000, in shares of £50 each. The Company also obtained power to raise their capital to the extent of £200,000, if necessary.

This line was opened to Moore's Junction, a distance of 32 miles, in August, 1852, and, with the Montreal and Lachine Railway, formed the connection between Montreal and Plattsburg, on the west side of Lake Champlain, and by ferry with the Rutland and Burlington Railway. This line is operated by the Grand Trunk Railway Company. Gauge, 4 ft. 8½ in.

CHIEF OFFICE—Point St. Charles, Prov. Quebec.

MONTREAL AND VERMONT JUNCTION RAILWAY.

This line of railway extends from St. Johns to St. Armand, a distance of 26 miles, and forms a connecting link between St. Albans and Montreal. The guage is 4 ft. 8½ inches. The road is operated by the Vermont Central Rail way Company.

CHIEF OFFICE—St. Albans, Vermont.

STANSTEAD, SHEFFORD AND CHAMBLY RAILWAY.

This line extends from' St. Johns to Waterloo, a distance of 42 miles. The first section was opened for traffic in January, 1860.

The cost of construction and equipment is over one million dollars. The road is now leased in perpetuity to the Vermont Central Railway Company. The account of revenue and expenditure does not appear separately.

CHIEF OFFICE—St. Albans, Vermont.

ERIE & NIAGARA RAILWAY.

(LEASED TO GREAT WESTERN RAILWAY COMPANY.)

A Company was incorporated to construct this line as early as April 16th, 1835, with a capital of £75,000. It was to run from some point on the River Welland to the Niagara River at or below Queenston, with power to extend to Lake Erie or to the Niagara River below Lake Erie, and from Queenston to Lake Ontario. The charter was amended in 1852, the capital being increased to £150,000, or £250,000 provided the line were extended to Lake Erie. At first a horse railway was constructed, but this was superseded by an iron road. The line, after being worked for a number of years, became involved, and operations were suspended; it was, however, reopened in 1867. The project was aided by subscriptions and loans as follows: town of Niagara, $60,000; town of Chippawa, $20,000; other loans, $220,000; total, $300,000. The road was opened to Chippawa—17 miles—in July, 1854, and afterwards to Niagara, 31 miles. The line is now operated to Fort Erie by the Great Western Company under a lease.

BUFFALO & LAKE HURON.

(LEASED IN PERPETUITY TO THE GRAND TRUNK.)

This line extends from Fort Erie, opposite Buffalo, on the Niagara River, to Goderich, on the east shore of Lake Huron, a distance of 162 miles. The road was opened from Fort Erie to Paris on the 1st November, 1856, 88 miles; from Paris to Stratford on the 22nd December, 1856, 33 miles; and from Stratford to Goderich, 28th June, 1858, 45 miles; total, 162 miles. Aid was granted to the undertaking by the municipalities along the line either in the shape of stock or bonuses, to the total amount of $878,000, as follows:

United Counties of Huron and Bruce	$300,000
County of Perth	200,000
Town of Stratford	100,000
" Paris	40,000
" Brantford	100,000
Township of Brantford	50,000
" Wainfleet	20,000
" Canborough	8,009
" Moulton and Sherbrooke	20,000
" Bertie	40,000
	$878,000

The amount expended on construction account, and for equipment, to 1st January, 1867, was $8,000,790. The capital account stood on the 30th June, 1869, as follows: bonds, £727,786 stg.; share capital authorized, £2,400,000. The amount of shares subscribed was £1,660,000; amount received on capital account, £1,775,071; amount expended, £1,701,056; balance unexpended, £74,015.

DIRECTORS (elected 1870).—Maxwell Hyslop Maxwell, Liverpool, Eng., chairman. Arthur Ashton, Liverpool; Samuel B. Healey, Liverpool; J. Johnson Stitt, Liverpool. Secretary—Thomas Short, Liverpool.

CHIEF OFFICE, 1 Great Winchester-street buildings, London, E. C.

PETERBORO' & HALIBURTON RAILWAY.

(PROJECTED.)

This Company was incorporated by an Act of the Ontario Legislature, passed on the 23rd January, 1869. The charter authorizes the building of a wooden or iron railway, of the 5 ft. 6 in. gauge, from the Town of Peterboro' to the town plot of Haliburton, in the County of Peterboro'. The capital authorized is $250,000, in 5000 shares of $50 each. So soon as $50,000 is subscribed, and 10 per cent. paid up, nine directors may be elected, and the Company fully organized. The road must be commenced within two years, and completed within four years from the date of the charter. The Town of Peterboro' has granted a bonus of $25,000 in aid of the road, and the Municipality composed of the united Townships of Dysart, Guilford, Dudley, Harburn, Harcourt and Burton, are also to extend aid to the undertaking. A liberal grant has, we are informed, been promised by the Ontario Government, out of the "Railway Fund." The manner in which the project will be carried out is yet uncertain, but it is anticipated that decisive action will be taken immediately.

DIRECTORS.—(Provisional, appointed in the charter)—P. Grover, John Carnegie, jr., George Read, Wm. A. Scott, Elias Burnham, W. H. Scott, James Stevenson, S. S. Peck, N. Kerchoffer, Francis Beamish, A. Trefusis, H. Williams, A. J. Cattanach, C. J. Bloomfield.

CHIEF OFFICE, Peterboro', Ont.

*EUROPEAN AND NORTH AMERICAN RAILWAY.

Fiscal Year ending 30th June.

Rolling Stock, &c.	1861.	1862.	1863.	1864.	1865.	1866.	1867.	1868.	1869.	1870.
No. of Locomotives....	14	14	14	14	14	14	14	14	14	15
" Passenger Cars.....	18	18	18	18	18	18	18	18	19	21
" Freight Box........	63	63	63	63	63	63	63	63	63	83
" Bag'e Mail & Exp.	4	4	4	4	4	4	4	4	4	7
" Platform............	105	105	105	105	105	105	105	115	115	115
Mileage Pas. Cars, 1st c.	....	132,620	137,862	144,985	127,398	122,801	....	159,708	167,682	174,886
" Freight Box...	... .	244,714	301,047	305,598	330,021	359,003	....	468,018	498,128	554,547
" Locomotives...	193,683	160,421	165,897	175,747	148,328	149,830	171,105	182,212	185,967	198,635
Passengers carried, No.	171,291	132,094	130,688	139,554	144,366	148,924	159,119	171,453	169,058	191,142
Freight, Tons...........	33,386	32,788	45,334	55,355	44,518	51,205	55,998	63,450	67,430	68,542
Revenue.	$	$	$	$	$	$	$	$	$	$
Passenger Traffic........	69,558 03	51,382 22	57,832 70	64,292 52	61,720 83	65,931 32	67,273 80	70,669 01	75,695 11	81,973 69
Freight "	47,700 72	46,784 53	61,388 78	71,999 74	64,349 96	72,685 64	76,271 01	87,970 24	97,089 80	103,322 51
Mails and Sundries.....	13,419 40	9,473 33	10,051 04	8,765 60	7,337 88	9,713 71	10,825 25	8,119 17	10,010 44	10,261 16
Gross Earnings..........	130,678 15	107,640 28	129,272 52	145,057 86	133,408 67	148,330 67	154,370 06	166,758 42	182,795 35	195,557 36
Expenditure.										
Locomotive Power......	36,415 39	28,562 68	28,319 75	33,691 99	33,422 68	31,933 86	36,535 35	39,453 54	42,485 88	44,224 10
Car Expenses........ ...	18,774 61	14,966 59	17,013 03	22,008 64	20,092 43	19,065 15	22,848 15	25,675 67	24,420 70	22,257 75
Maint'e Way & Works	19,464 60	22,931 98	24,471 83	26,295 04	24,239 74	27,195 17	34,642 06	44,011 00	33,756 78	40,101 53
Station Expenses..	....	....	....	....	....	....	.. ..	....	16,544 43	21,596 88
General Charges.........	19,590 92	21,173 48	18,729 68	21,634 45	17,152 11	18,376 03	20,270 12	22,544 76	8,941 92	11,503 73
Total...............	94,245 52	87,634 73	88,534 29	103,630 12	94,906 96	96,570 21	114,295 68	131,684 97	126,149 71	139,683 99
Net Revenue.......	36,432 63	20,005 55	40,738 23	41,427 74	38,501 71	51,760 46	40,074 38	35,073 45	56,645 64	55,873 37
Per centage of Working expenses to gross recpts	72 12	81 04	68 48	71 44	71 19	65 10	75 45	78 96	69 01	71 42

*Omitted at page 120.

CANADIAN RAILWAY TRAFFIC.

RECEIPTS OF FOURTEEN LINES.

Railways.	1866.		1867.		1868.		1869.		1870.	
	Miles open.	Total Receipts.	Miles open.	Total Receipts.	Miles open.	Total Receipts.	Miles open.	Total Receipts.	Miles open.	Total Receipts.
		$		$		$		$		$
Grand Trunk	1,377	6,688,980	1,377	6,596,966	1,377	6,906,023	1,377	7,386,684	1,377	7,[illegible]
Great Western	345	3,754,402	349½	3,725,160	351½	3,734,392	351½	3,909,[illegible]	351½	4,067,[illegible]
London & Port Stanley	26	42,974	25	42,791	24½	41,704	24½	42,[illegible]	[illegible]	[illegible]
Welland	25	105,946	25	68,615	25	77,482	25	84,294	25	78,704
Northern	94	512,874	94	561,370	96	590,670	96	671,876	[illegible]	7[illegible],467
Midland of Canada	66	....	56	253,477	56	[illegible]	56	[illegible],561	[illegible]	342,1[illegible]
Cobourg, Peterboro' & Marmora	....	....	28	21,078	28	18,341	22	32,[illegible]	[illegible]	44,[illegible]
Brockville & Ottawa	86	116,594	86	126,660	[illegible]	178,373	86	164,945	[illegible]	[illegible]
Canada Central	....	....	....	....	....	....	....	....	[illegible]	*7,[illegible]
St. Lawrence & Ottawa	54	104,490	54	106,515	54	117,470	54	12[illegible]	54	141,[illegible]
St. Lawrence & Industry	12	6,066	12	7,860	12	....	12	6,273	12	6,817
New Brunswick & Canada	....	65,320	....	79,781	....	108,348	126	104,[illegible]	12[illegible]	141,[illegible]
European & North American	88	89,738	107	96,777	12[illegible]	106,780	106	102,796	127	[illegible]
Nova Scotia	59	190,730	....	756,[illegible]	16[illegible]	263,994	[illegible]	373,[illegible]	[illegible]	27[illegible]
Windsor & Annapolis	....	....	....	....	52	....	....	....	....	....
Total	2,318	11,147,894	2,215	12,321,069	2,3[illegible]	12,[illegible]	2,[illegible]	13,[illegible]	2,[illegible]	13,717,477

* 3 Months.

BAINES'

REVERSIBLE ROLLS.

Important Facts for Railway Companies.

One mile Steel Rail, at 70 lbs. per yard, or 123 tons, 400 lbs. at $70		$8,624 00
One mile of new Iron Rail, at 70 lbs. per yard, or 123 tons 400 lbs, at $45		5,544 00
One mile of re-rolled Iron Rails, at 70 lbs. per yard, or 123 tons 400 lbs., converted into new rails, at $35.00		4,312 00
One mile of old rails, at 70 lbs. per yard, or 123 tons. 400 lbs.; re-rolled in damaged parts by Reversible Rolls, average cost per ton, $3.50 at the most	$431 20	
Royalty per mile	50 00	
Interest on first outlay for Mill and Machinery, per mile	6 60	
		$487 80

Railway Companies anxious to re-lay their lines, say with 10 miles of steel rail per annum, could repair 10 miles of their best old iron rails, by Reversible Rolls, at a cost of $4,878, rather than re-roll the same quantity by the old fashioned plan, at a cost of $4,312.00 per mile.

One set of Rolls, with two furnaces, is capable of repairing per annum each rail in every 45 miles of road, each rail having on two patches, not exceeding 3 *ft.* 6 *in.* in length.

Railway Companies adopting the Rolls, say on a 300 mile road and repairing only about 8 miles the first year, would effect a saving sufficiently large to pay for patent right, cost of Mill and Machinery.

Railway Companies determined to continue using Iron Rails would effect an enormous saving by using the Reversible Rolls.

Railway Companies having Reversible Rolls, could purchase good old rails, and repair same with great profit to themselves.

By Baines' Improved Rolls, and new shape of Patch, the Scarf is made, without doubt, the strongest part of the Rail, and that, too, merely upon its passing through the Rolls at proper welding heat, not in any measure depending upon the care and attention of workmen.

For further information apply to

HUGH BAINES,

TORONTO, CANADA.

or to

MATT. TAYLOR,

21 *Nassau Street, New York, U. S.*

Toronto, 9th Nov., 1870.

Baldwin Locomotive Works.

M. BAIRD & Co., Philadelphia,

Manufacturers of

LOCOMOTIVE ENGINES!

OF ALL SIZES AND DESCRIPTIONS,

Especially adapted to every variety of Railroad Service,

And to the economical use of Wood, Coke, Bituminous and Anthracite Coal, as Fuel.

SPECIAL PATTERNS OF SMALL LOCOMOTIVES

Designed expressly for Narrow Gauge Railroads.

All Work accurately fitted to Gauges,

AND THOROUGHLY INTERCHANGEABLE.

Plan, Material, Workmanship, Finish and Efficiency fully guaranteed.

For full particulars address

M. Baird, Edw. H. Williams,
Geo. Burnham, Wm. P. Henszey,
Chas. T. Parry, Edw. Longstreth.

M. BAIRD & Co.,
PHILADELPHIA, PENN.

Canadian Engine & Machinery Co.,

KINGSTON, ONT.

The works of this Company are completely and fully equipped with the most approved of modern Tools and Machinery, for building

RAILWAY LOCOMOTIVES,

RAILWAY CARS,

Railway Snow Ploughs,

IRON RAILWAY BRIDGES, &C., &C.

We have already built Locomotives for all the principal Railways in the Dominion, as will be seen by the following list, viz:

Grand Trunk Railway,
Great Western Railway,
Nova Scotian Railway,
Northern Railway,
St. Lawrence & Ottawa Railway,
Brockville & Ottawa Railway,
Midland Railway,
Cobourg & Peterboro' Railway,
Toronto & Nipissing Railway.

And we are now building for the Intercolonial Railway.

For further particulars apply to the undersigned,

CHARLES GILBERT,
Secretary & Treasurer,
KINGSTON.

R. J. REEKIE,
Managing Director,
Great St. James Street, Montreal.

THE

HINKLEY & WILLIAMS WORKS,

552 Harrison Avenue, Boston, Mass.

ADAMS AYER, Pres. F. L. BULLARD, Treas. H. L. LEACH, Supt.

MANUFACTURERS OF

LOCOMOTIVE ENGINES,

BOILERS, TANKS,

IRON AND BRASS CASTINGS, &C.

Of the most improved description, and of the best material and workmanship. They are prepared with patterns, and can furnish at short notice, engines of the following general descriptions and dimensions.

Eight-Wheeled Locomotives, with Four Drivers and Truck.

Weight, 22 tons. Driver, 4½, 5 or 5½ diameter.	Fire Box, 60 in. long.	Cylinder, 16 or 17 × 24.
" 30 " " " "	" 60 "	" 16 or 17 × 24.
" 28 " " " "	" 54 "	" 16 × 22 or 24.
" 26 " " " "	" 50 "	" 15 or 16 × 22.
" 24 " " " "	" 48 "	" 14 or 15 × 22.
" 22 " " " "	" 42 "	" 13 or 14 × 22.

Eight-Wheeled Locomotives, with Six Drivers and Two-Wheeled Truck

Weight, 37 tons. | Driver, 4½ or 5 ft. diameter. | Fire Box, 66 in. long. | Cylinder, 18 × 22.

Four-Wheeled Switching Locomotives.

Weight, 21 tons.	Drivers, 50 inches diameter.	Fire Box, 42 in. long.	Cylinder 14 × 22.
" 18 "	" " "	" "	" 13 or 14 × 22.

They will also contract to build Locomotives to specifications of any design, or will modify the above proportions to suit purchasers.

ESTABLISHED 1868.

CANAL
IRON NAIL & SPIKE WORKS,
MONTREAL.

PECK, BENNY & CO.,

MANUFACTURERS OF

Railroad Spikes,
Ship Spikes,

AND ALL DESCRIPTIONS OF

Cut Nails, Pressed, Clinch and Slate Nails.

Office, 391 St. Paul Street. Works, 61 Mill Street.

THE

Detroit Bridge & Iron Works,

OF DETROIT, MICH.

MANUFACTURERS OF

Bridges, Roofs, Turn Tables, &c.

OF IRON OR WOOD.

PIVOT DRAW BRIDGES & FIXED SPANS of any required Length.

WE BUILD

The Bollman Truss,

The Fink Truss,

The Pratt Truss,

The Triangular Truss or

Warren Girder,

Boiler Plate or Trussed Girders,

OR ANY OTHER APPROVED STYLE THAT MAY BE DESIRED.

ALSO

COMBINED WOOD AND IRON BRIDGES,

With Lower Cords of Wrought Iron,

Iron Roof Trusses & Frames.

Every variety of heavy or light Forgings, Shafts, Axles, Connecting Rods, Cranks, etc. etc. Also all kinds of Castings, Fish Plates, and all kinds of Bolts, Nuts and Rivets for Bridge or Building purposes.

ALSO

STEAM ENGINES

AND

Mill and Machine Work Generally.

Plans and Specifications furnished, with Estimates, if required.

Our Bridge and Roof Work is in large use in almost all the important Railways in the North-West, to the officers of any one of which we confidently refer.

Office and Works: At Corner of Foundry St. and M.C.R.R.

DETROIT, MICH.

WILLIAM C. COLBURN,
Secretary and Treasurer.

WILLARD S. POPE,
Engineer.

W. J. M. JONES,

(SUCCESSOR TO ALFRED BROWN)

DEALER IN

Railway Supplies,

20 ST. SACRAMENT STREET.

MONTREAL.

CANADIAN RAILROAD

LAMP MANUFACTORY,

50 Queen Street West. Toronto.

JOHN BOXALL,

PROPRIETOR.

Manufacturer of all kinds of Lamps

LOCOMOTIVE HEAD LAMPS & BURNERS,

TAIL, SWITCH, GAUGE AND SIGNAL LAMPS, SPERM & COAL OIL HAND LAMPS.

Coal and Wood Stoves of every description. Hot Air Furnaces, &c.

RAILWAY & TELEGRAPH MACHINERY & SUPPLIES,

L. C. TILLOTSON & CO.,

MANUFACTURERS OF

Railway Car Findings,

IMPORTERS AND DEALERS IN

Rubber Springs, Plushes, Gilt Mouldings, Seat Springs, Car Ventilators, Locks, Spittoons and Lamps,

UPHOLSTERERS' MATERIALS, LANTERNS, OILS,

Head Lights, Steam Gauges, Brass, Copper and Iron Flues, Nuts, Washers, Bolts, Wrenches, Files, Locomotive Gauge, Steel Tire, Belting, Hose, Cotton Waste, Steam Packing, Hydraulic Jacks, Conductors' Punches and every description of Engine and Car Supplies.

Sole Agent in the U. S. for Richard Johnson & Nephew, Wire Manuf'rs, Manchester, Eng.

No. 8 Dey Street, New York.

CANADA FILE WORKS.

English Oak-Tanned Leather

BELTING,

AND

Lacing Leather Works,

Railway, Steamboat and General Engineering Supplies, Railway Steel and Rubber Springs, Cast and Spring Steel, Rubber Belting, Hose and Packing, Patent Anti-friction Steam and Pump Packing, Rivets, Nuts, Washers, Babbitt Metal, Steam Gauges, Oxide of Iron, Paints, &c., &c., &c.

DIVING APPARATUS COMPLETE.

PRINTERS' PRESS BLANKETS.

W. L. KINMOND & Co.

Consulting Mechanical Engineers, MONTREAL.

INCORPORATED APRIL, [illegible]

TORONTO CAR WHEEL COMP'Y

CAR,  TENDER

AND

LOCOMOTIVE WHEELS!

MANUFACTURED FROM

THE BEST COLD BLAST CHARCOAL IRON.

WORKS:

Corner of Esplanade and Alfred Streets,

(OPP. THE QUEEN'S HOTEL), · **Toronto.**

JOHN GARTSHORE MANAGER.

www.ingramcontent.com/pod-product-compliance
Lightning Source LLC
Chambersburg PA
CBHW020616030726
47497CB00007B/2275
9783744724203